全国高等农林院校"十三五"规划教材

A Course in Marine English Translation

海洋英语翻译教程

周永模　主编

中国农业出版社

图书在版编目（CIP）数据

海洋英语翻译教程/周永模主编．—北京：中国农业出版社，2018.1（2020.8 重印）
全国高等农林院校“十三五”规划教材
ISBN 978-7-109-23859-6

Ⅰ.①海…　Ⅱ.①周…　Ⅲ.①海洋学－英语－翻译－高等学校－教材　Ⅳ.①P7

中国版本图书馆 CIP 数据核字（2018）第 010799 号

中国农业出版社出版
（北京市朝阳区麦子店街 18 号楼）
（邮政编码 100125）
责任编辑　龙永志　马顗晨
文字编辑　蔡　菲

北京印刷一厂印刷　　新华书店北京发行所发行
2018 年 1 月第 1 版　　2020 年 8 月北京第 2 次印刷

开本：720mm×960mm 1/16　　印张：18.75
字数：335 千字
定价：39.50 元

编审人员名单

主　编： 周永模

副主编： 陈　橙　梅进丽　马百亮

编　者：（按姓氏笔画排序）

马百亮　陈　橙　周永模　梅进丽

主　审： 戴小杰

前 言

21世纪以来，我国海洋经济已经逐渐形成了一个由海洋渔业、海洋交通运输业、海洋油气业、滨海和海岛旅游业、海洋盐业、沿海造船业等构成的体系化产业，不但海洋经济总量有了较大的发展，而且有些海洋产业甚至居于世界沿海国家前列。但相比我国经济发展规模以及世界海洋开发总体水平，我国海洋资源开发还比较落后，存在着观念滞后、开发技术不足、结构不合理、海洋生态恶化、海洋资源权益之争繁多等问题。要想解决这些问题，需要我国年轻一代的海洋资源开发者，特别是有志于海洋学科以及海洋相关学科发展的专业人士，能够以一种全新的姿态去关注海洋和认识海洋，以实现海洋资源的可持续开发和利用。

因此，海洋知识的普及，海洋文化的挖掘，尤其是基于海洋的国际间的交流和合作等工作亟待开展。而这些工作都离不开高素质的翻译人才。这类翻译人才不仅应具备扎实的语言基本功和较强的英语综合应用能力，同时还应兼具一定的宽泛的海洋学科专业知识。在此背景之下，我们策划并组织编写了这本《海洋英语翻译教程》，以期开拓翻译类教材编写的新视角。将对翻译基本理论和技巧的探讨与特定学科相对接，以提升翻译技能学习和培养的实效性。

本教材由导论、十二个主题单元和附录等三部分构成，每个单元聚焦不同的海洋主题，探讨与介绍不同翻译技巧。每个单元分别包括以下几部分内容：

第一部分：词汇预习。这部分主要包括本单元知识导读、翻译技巧例句解析和课后练习中出现的词汇以及与主题相关的海洋学科高频英语词汇，为课文学习做热身准备。

第二部分：课文导读。本教材共设海洋演变史、海洋生物资源、海洋矿产资源、海洋污染、海洋灾难、海洋保护、海岛旅游、海洋法律法规、海洋工程、航海运输、海洋经济和海洋管理等十二个与海洋相关的主题，较为全面地介绍了海洋学科基础知识和海洋研究中一些热点问题。在每个单元的主题范围内选取一篇约2 000词的具有代表性的文章，作为相关主题的背景知识导读内容。本部分旨在通过提供这些兼顾学术性和科普性的导读材料，帮助学生拓宽海洋科普视野和思路，形成对海洋科学整体的认识，为其日后进行本学科翻译工作提供知识储备。同时，导读材料也可为各单元特定翻译技巧的集中探讨以及课后练习提供例句材料。

第三部分：翻译技巧介绍。这是本教材的核心部分，主要探讨和例证一些常见和常用的翻译理论和技巧，每个单元聚焦一种翻译技巧。全书分别从理解与表达，直译、意译与音译，词类转换，增译，省译，正译与反译，长句与复杂句，语态转换，海洋英语术语的构成与翻译，词序调整，分译，校对等十二个方面展开论述，并辅以从海洋英语文献中精选的若干例句，以帮助读者更加深刻地理解这些翻译技巧在海洋英汉翻译实践中的具体运用。所有例句均与海洋科学相关，但视具体情况，或取材于背景知识导读部分，或取材于其他英语文献资料。

第四部分：课后练习。每单元的练习均由单句翻译和短文翻译两部分组成，其中单句翻译包括英译汉和汉译英各15个句子，短文翻译则包括英译汉和汉译英短文各一篇，篇幅为500～800词。

与国内已经出版的诸多翻译教材相比，本教材具有如下几个特点：

1. 多用性。本教材既可以作为涉海行业高校翻译专业研究生的必修或选修教材，又可以作为海洋科学或相关学科或专业研究生英语学习的拓展教材。除此之外，具有一定英语基础，且对海洋科学知识感兴趣的英语爱好者还可将本教材作为拓宽海洋科技知识和提高英语翻译技能的辅助学习材料。

2. 规范性。本教材中的例句和短文语料均节选或选编于近年来英语国家出版的海洋专业教科书、海洋科普知识网站或国内英语期刊选载的海洋英语读物，文章用词严谨，语言规范，内容上兼具学术性、知识性和趣味性的特点。

3. 实用性。在论述翻译理论和技巧时，本教材努力吸收各家之长，既保证为读者提供较为全面的传统译学基本知识和技能，又十分注重所述翻译技巧的学科性和实用性。考虑到翻译学习需要大量实践的特点，本教材对翻译基本理论、基本知识和基本技巧等的讲解力求简明扼要，对练习材料的编选给予了特别的重视。读者可通过书中提供的大量练习，巩固有关理论和知识，提高翻译技能。书后附录部分附有各单元练习的参考答案，以便于读者进行翻译实践和自学。

本教材总策划为周永模。除编审人员名单中所列人员，参加本教材审读的还有上海海洋大学爱恩学院外籍教师Sernan先生和美国大峡谷州立大学（Grand Valley State University）的英文写作课教师Craig Hulst博士。本教材的问世凝聚了多人的辛劳和汗水，上海海洋大学外国语学院相关教师和海洋科学学院戴小杰教授在选材上提供了帮助，并提出了许多宝贵和中肯的修改意见，在此一并表示感谢。

因编者水平有限，时间仓促，缺乏经验，且对有关的专门科学领域知识储备不足，选材或多有遗漏，或有失过泛。疏漏在所难免，祈请专家和读者不吝教正。

编　者

2017年10月

CONTENTS

导　论

我国科技英语翻译的发展已有一千多年的历史。较大规模的发展，是从明末清初西方传教士与中国士大夫大量翻译西方科学技术著作开始的。中华人民共和国成立后，兴起了有组织的科技英语翻译及其研究。到了21世纪，高科技机构的建立使科技英语翻译变得更加专业化，翻译范围不断扩大。在当今时代，经济迅猛发展，网络技术日新月异，国际合作交流广泛，一个国家能否以最快的速度翻译获取最先进的科技信息，并将本国科技的最新发展翻译介绍给其他国家，是衡量国家发达与否的重要尺度之一。因此，科技英语在国际学术和生产技术交流中的重要性日益凸显，这对高等学校培养既懂英语、又懂专业技术知识的复合型人才提出了较高的要求。

科技英语涵盖的范围很广，通常包括自然科学和工程技术方面的著作、论文、研究报告、学术演讲、产品指南等文本中使用的英语。科技英语翻译的主旨是传递科技信息、传播科技知识、进行科技交流，确保科技信息传递的真实性。从科技英语翻译的过程来看，主要包括分析、理解、转换和表达几个重要步骤。分析是指对科技语体的长句、复杂句等认真剖析，厘清句子各成分之间的关系。理解是指准确理解原文所传达的信息，尤其是科技专有名词术语等。转换是运用多种翻译策略和方法，将原文中的科技信息转化为译文中的等值信息。表达是将这一等值信息用符合译入语的方式准确无误地进行再现重组。

科技英语翻译要求精确性第一，规范性第一，绝不可言不达意。任何笼统含糊、生硬牵强的表达都不应出现在科技英语翻译中，不然会影响甚至破坏科技信息的有效传递。

首先，科技英语翻译中选词用字要符合行规，不说外行话。在原文意义的理解和译文的选词用字方面，要特别注意词语在特定上下文中的内涵和外延意义。要确定一个词语的具体意义，必须将其置于特定的上下文中认真推敲，甚至根据整个语篇予以确定。实际上，科技英语最难理解和翻译的不是专业词汇，而是一些半科技动词、副词和形容词，特别是一些短语动词，常常需要译者了解多义词的每一个含义及当时的语境，并能根据专业知识加以适当的判断来选择词义。相对来讲，科技英语的专业词汇词义专一，一词多义较少。当有新词出现时，译者要懂得利用构词法理解词义并准确恰当地译出。

其次，科技英语翻译要特别注意分析文章的整体思路和发展脉络，使译文达到文从字顺、意脉通畅。在科技英语翻译中，英汉两种语言不同的行文习惯和结构差异要求译者在行文用字时倍加谨慎。如果只是单纯强调传递原文的真实信息，而忽视信息传递的效果，则会导致译文生硬牵强、文字不通顺、信息不明确，就谈不上信息传递的真实和准确。所以，在翻译时，除注重遣词用字上的精准与规范外，还应尽量避免那些佶屈聱牙的翻译腔，使译文通顺自然，让读者易于接受。

海洋英语属于科技英语的一个分类，也可以称之为海洋科技英语。海洋科技英语以英语为载体，传递海洋领域的研究成果及研究走向，汇集了大量的海洋词汇与表达方式。从内容上说，海洋科技英语主要包括海洋生物、海洋矿产、海洋污染、海洋保护、海洋旅游、海洋工程、海洋运输、海洋管理等，不一而足。海洋英语所涉及的专业领域广、知识丰富、信息量大，因此译者在进行海洋英语翻译时，除了需要具备过硬的中英文语言能力，更需要拥有跨学科的综合知识，尤其是海洋类专业领域知识。只有不断吸收和丰富科学文化和海洋专业技术知识，才能理解海洋英语文献中的专业术语、基本概念、语篇信息等。

良好的科学素养和丰富的海洋专业知识，是准确翻译的前提。在此基础上，译者还需要熟悉翻译理论，将各种翻译策略和翻译技巧灵活运用到海洋科技英语翻译的实践中。与科技英语翻译的基本特点一致，海洋科技英语翻译有许多专有名词和固定表达，译者不能擅自创译，否则会造成读者的误解，因此在表述专有名词时必须参照海洋学术语的国家标准，或与国外学界规范表达一致。海洋科技英语翻译必须尽量做到专业化和专门化，这是科技英语翻译的实质。因此合格的海洋科技英语翻译工作者应该熟知相关的海洋专业知识和文体模式。同时，要注意英汉两种语言之间的差异，注意译文的准确、简洁、通顺。在准确理解原文的基础上，综合运用各种翻译方法，包括直译、意译、音译、转换（各种词类如名词、动词、形容词、副词、介词和连词等的相互转换；各种句子成分如主语、宾语、谓语、定语、表语、状语等的相互转换等）、增译（增补语义上和修辞上需要的词；增补原文中的省略成分和隐含含义等）与减译（冠词、代词、介词、连词、动词、同位语和复合名词等的减译）、正译与反译（各种句子成分和结构在顺序上正着译和反着译）、分译与合译（把长句进行拆分，把短句进行合并）等，最大限度地缩小目的语和原语的差异。

在海洋英语翻译的理解和表达中，如果仅仅对句子的表层词汇意义进行理解和拼缀，而没有深入句子的深层关系进行剖析，就容易出现误解和误译。因此，在翻译时，一定要有意识地识别句子中的形态词（名词、动词、形容词和副词等）和结构词（介词、连词、冠词、关系代词和关系副词等），了解句子的基本句型、成分和语法关系，进而深入了解句子语言成分的概念范畴之间的关系。在遇到长句时，尤其需要通过识别形态词，突显主、谓、宾、表等主干成分，厘清主干成分与次要成分之间的逻辑关系和修饰关系，然后通过适当的方法翻译出来。

虽然科技英语翻译已有一千多年的历史，海洋科技英语翻译却是一个新生事物。为应对新形势下国家海洋强国的发展战略，众多涉海领域均需要既具备海洋学科专业背景，又具有良好的英语应用能力的国际化复合型人才。因此，海洋科技英语的翻译就显得尤为重要。本教材涵盖了有关海洋科技的主要方面。在教材的宏观组织、课文的选编、练习的设计上，一方面重视基本理论，力求向学习者提供海洋科技英语翻译的基本理论知识，旨在培养专业学习者所必备的专业技能；另一方面，强调海洋科技英语基本技能的训练以及文化意识的培养，旨在从理论层面上进一步强化训练学生良好的语言技能，使他们具有从事海洋科技英语翻译实践工作的业务水平与基本素质。

总体来说，科技英语翻译应该坚持先理解到位，后着手表达。在海洋科技领域，更是如此。面对科技前沿的海洋知识，译者切不可因一字之缪，造成不可弥补的损失。加强语言功底的修炼和海洋专业知识的训练，并且着力培养自身文化素养，体会不同文化之间的异同，才能架设好翻译这座桥梁，以期为我国的海洋科技研究、海洋经济发展以及海洋战略决策搭建起一条翻译界的“丝绸之路”。

Unit One The History of the Oceans

第一单元 海洋演变史

Ⅰ. Warm-up for Theme-related Words/Expressions 词汇预习

atmosphere *n.* 大气，大气层，大气圈
basalt *n.* 玄武岩
boulders *n.* 卵石，圆石；巨砾，冰砾；漂砾
carbon dioxide 二氧化碳
coalesce *v.* 使联合，使合并
coastline *n.* 海岸线
comets *n.* 彗星
continent *n.* 大陆；洲
the Continent 欧洲大陆
continental crust 陆壳
continental rocks 大陆岩石
core *n.* 地核，核心
crust *n.* 地壳；外壳
crustal rocks 地壳岩石
degas *v.* 排气，排除煤气
deep depression 深凹区，深坳陷
gaseous state 气（体状）态
"Goldilocks" planet 适合（人类）居住的行星
granite *n.* 花岗岩
gravitational attraction 地球（地心、万有）引力
greenhouse *n.* 温室，玻璃暖房
hydrosphere *n.* 水圈，水界，水气
ice cap 冰帽，冰冠
lithosphere *n.* 岩石圈，地壳
magnetic anomaly 磁异常
marine biology 海洋生物学
mantle *n.* 地幔
Mars *n.* 火星
meteorite *n.* 陨星；陨石；陨铁；流星
methane *n.* 甲烷，沼气
molten rock 熔融岩石
nickel *n.* 镍
ocean crust 洋壳
organism *n.* 有机体；生物体；有机组织
planet *n.* 行星
pangaea *n.* 泛大陆，盘古大陆
planetary bodies 行星体
planetesimal *n.* 小星体 *a.* 星子的，星子组成的
plate tectonics 板块构造（论）
Pluto *n.* 冥王星
radioactive elements 放射性元素

rotating disk 旋转圆盘
sea floor 海底；海床
solar radiation 太阳辐射
subduction *n.* 消减，潜没，俯冲（作用）（指地壳的板块沉到另一板块之下）
submarine mountains and valleys 洋底山脉与山谷
sulfur *n.* 硫黄（色） *v.* 用硫黄处理
temperature gradient 温度变化率
the East Pacific Rise 东太平洋（海隆）海岭
the mid-oceanic ridge system 大洋中脊（隆）系统
the solar nebula 太阳星云
the solar system 太阳系
tidal rhythm 潮汐韵律（层）
timescale *n.* 时间表［尺度］，时标，时间量程
transform faults 转换断层
habitat *n.* 栖息地，产地
trench *n.* 海沟
underlying rock 基岩，下垫岩石，下伏岩石
UV *abbr.* 紫外线（的）（＝ultraviolet）
Venus *n.* 金星
volatiles *n.* 挥发物
volcanic islands 火山岛
volcanic rocks 火山岩
volcanism *n.* 火山活动，火山作用，火山现象

Ⅱ. Lead-in & Reading 课文导读

How Did the Oceans Form?

(1) Seventy percent of the Earth's surface is covered with water. Where did it come from? How did it make its way from the deep Earth to the surface of the planet? And most importantly, why has liquid water persisted on Earth when it doesn't appear to have done so elsewhere in the solar system? The full story takes us back to the formation of our "water planet".

The Birth of the Solar System

(2) Everything in our solar system—the Sun and the planets—formed about 4.55 billion years ago from a rotating disk of gas, dust, and ice called the solar nebula. Matter in the nebula came together to form planetary bodies of different sizes and composition, with the Sun at the center. Rocky planets formed closest to the Sun, the gas giants (气体巨行星) farther out, and icy bodies like Pluto farther still. This is because the nebula cloud was hotter toward the center, just like the solar system today.

(3) This temperature gradient influenced the composition of these planetary bodies, including the concentration of water. Lighter elements and compounds known as volatiles (things like water, carbon dioxide, and methane, which are stable as gas in our atmosphere) accumulated further away from the Sun, concentrating in bodies like the gas planets. The rocky planets, like Venus, Earth, and Mars, formed where it may have been too hot for the water and other volatiles that form our hydrosphere and atmosphere to be present in high concentrations.

Earth Takes Shape

(4) Earth formed as dust, rocks, and planetesimals (kilometer-sized boulders) collided and combined to create something just a bit smaller than the present planet. These collisions, combined with the decay of radioactive elements, created an extremely hot planet. Geologists believe that soon after Earth formed, its surface started to melt, forming an ocean of molten rock approximately 400 kilometers deep. Earth was too hot to sustain water on its surface, which boiled and evaporated into the blistering atmosphere.

(5) That's when a process called differentiation began. Heavier elements sank to form the Earth's core, which is rich in iron and nickel. The remaining elements went into forming the rocky mantle that makes up most of the

planet's volume (84%). An outer crust, like the skin of an onion, composed of light elements, was extracted from the mantle as volcanic rocks. Our atmosphere (the blanket of gas that surrounds the Earth and is kept in place by gravitational attraction) is composed of the lightest elements of all.

(6) This differentiation—the separation of the planet into layers that are less dense as they extend outward from the core—happened very quickly by geological standards: within the first 100 million years after the Earth's formation. This process is ongoing and is what makes the planet so dynamic. Plate tectonics, for example, occurs because the planet continues to get rid of its heat.

The Role of the Moon

(7) The Moon probably formed some 50 million years after the Earth, which is when scientists think a Mars-sized body collided with our planet. This theory holds that some mass was added to the Earth, but the impact caused most of the material (including a lot of material from Earth) to fly off and get trapped in the Earth's gravitational field. This stuff then coalesced to form the Moon in only a few hundred years. The Moon's gravitational pull on the Earth causes the Earth to expand and contract slightly, and causes the oceans to move in tidal rhythm.

Where Did Earth's Water Come From?

(8) There are two dominant theories:

(9) The inside-out model proposes that the Earth formed with trace amounts of water structurally bonded to the minerals in the mantle. This water makes its way to the Earth's surface through volcanic processes. The outside-in model proposes that the Earth formed without water, which came with other volatiles from the meteorites or comets that bombarded the young planet. This water was probably mixed into the upper layers of the Earth and was later brought to the surface through volcanism.

(10) Neither model is completely satisfactory, but most scientists support the first. Comets may have given the Earth a little bit of its water, and possibly as much as 20%, but nowhere near enough to fill the oceans. Meteorites, on the other hand, appear to be the building blocks of our planet

because they've been found to have a composition similar to that of the early Earth.

(11) Studies of meteorites allow scientists to estimate that as much as 0.5% of the weight of the Earth is made up of water. That may not sound like a lot, but considering how big the planet is, it's more than enough to fill the world's oceans. The Earth's mantle is estimated to contain between three to six times as much water as in the oceans, so it's perfectly feasible that our surface water came from inside the Earth.

How Did the Hydrosphere and Atmosphere Form?

(12) Earth scientists believe that during the first few hundred million years after the solar system formed, while the Earth was differentiating into a core, mantle, and crust, the planet started degassing ("burping" volatiles) through volcanic activity. Volatiles that were trapped deep in Earth were released when the rocks that contained them melted and were erupted from volcanoes. Carbon dioxide, methane, sulfur, and other volatiles stayed in their gaseous state to make up the early atmosphere. Erupted water vapor, on the other hand, largely condensed to form the early ocean. Today, volcanism continues to supply the atmosphere and hydrosphere with many of the same gases—think of the white plumes[1] that steam from active volcanoes—but at a much lower level than when the early Earth was releasing so much heat.

(13) The new atmosphere that formed was steamy and sizzling, holding in heat like a greenhouse and obscuring any direct view of the Sun. It would have been something like Venus today, where the dense atmosphere keeps the planet at a scorching 400 ℃ (752 ℉). While our Earth's ocean and atmosphere evolved into a life-nurturing system, our neighboring planets fared less well. When they formed, Mars and Venus probably contained concentrations of water similar to Earth's, but they didn't hold onto it. Because Venus is so close to the Sun and has a thick, insulating atmosphere, most of its water boiled off. Small in mass, Mars lacked the gravity to hold onto its volatiles, including liquid water. Furthermore, a thin atmosphere and a significant distance from the Sun combine to make the planet cold. The poles of Mars are believed to contain frozen water and carbon dioxide ice caps that extend underground, and recent Rover expeditions yielded evidence that liquid

water once ran on its red surface.

Things Cool Off

(14) Over millions of years, Earth's thick steam atmosphere slowly cooled to the point where water was stable as liquid. Clouds formed and the atmosphere rained on the oceans—probably one very shallow ocean covering the planet's surface, since the continents likely did not yet exist. (Since no geologic record exists for the first 150 million years of the Earth's history, we can only speculate.) By 4.2 billion years ago, the age of some of the oldest rocks, we know there was liquid water on the Earth's surface, and that the atmosphere was never so hot that it turned to steam nor so cold that it all froze.

Why Has Water Persisted on Earth?

(15) Earth may be considered a "Goldilocks" planet[2]: "just right" for liquid water, and thus for life, which requires it. Many factors play a part:

- Earth is just the right distance from the Sun. The atmosphere and the ocean regulate the planet's temperature, keeping it relatively constant and in a range where liquid water is stable.
- Earth is big enough for its gravity to hold the atmosphere in place.
- Both the atmosphere and the Earth's magnetic field protect the Earth's surface from harmful UV[3] and solar radiation from the Sun.

(16) Is the volume of water on Earth constant? Trace amounts of water are lost from the atmosphere into space. However, trace amounts are also added from the meteorites that are constantly bombarding the atmosphere. More water is added than lost.

(17) All these things contribute to an exquisite dynamic system that brought the planet's oceans into being and protects the organisms that inhabit them.

Notes:

[1] white plumes 白色的羽流（流体力学专业用语），是指一柱流体在另一种流体中移动（a column of one fluid moving through another），即活火山喷发时产生的气体在大气圈中移动。

[2] "Goldilocks" planet 适合居住的星球。Goldilocks 一般指金凤花姑娘。金凤花姑

娘是美国传统童话中的一个角色，由于她喜欢不冷不热的粥，不软不硬的椅子，总之是“刚刚好”的东西，所以后来美国人常用金凤花姑娘来形容“刚刚好”。

[3] UV 紫外线（的），即英文单词 ultraviolet ray 或 ultraviolet radiation 的简称。紫外线是由德国科学家里特发现的，是电磁波谱中波长为 100～400 nm 辐射的总称。紫外线的波长越短，对人类皮肤危害越大。

Ⅲ. Teaching Focus (1): Comprehension and Reproduction 翻译技巧（1）：理解与表达

任何一门语言的翻译都离不开对原语的理解和对译语的表达这两个步骤，其中，理解是翻译的前提和基础，表达则是翻译的关键和结果。理解是译者运用自己的语言知识、专业知识、文化素养以及对世界的感知与认识，对原文进行信息解读、逻辑推理，步步深入，从而确定语义的过程；而表达则要求译者在忠实于原语信息内容的基础上，灵活运用各种翻译技巧和方法，发挥译语的功力和表达优势，在字词句的处理和语体风格的把握等层面，都能忠实通顺地再现原文。

1. 理解 “译路漫漫，始于理解。”理解通常可分为宏观和微观两个层面。

从宏观上来看，理解包括对原作的题材、语篇类型、话题类别、主题思想、表达风格、写作目的和交际对象等重要语境因素的把握和定夺。对原作内容的宏观理解在整个语篇翻译中具有一定的指导作用，也对原语的微观理解与表达有一定的帮助。例如，和其他行业领域英语一样，海洋英语文献作为一种文体，以说明文（expository writing）见多，而说明文又具有简洁性、准确性、客观性、科学性、逻辑性和知识性等特点，以“作者说”和“让人明”为主要表达方式来解说事物、阐明事理和剖析现象，从而给人以知识。说明文通过对实体事物的解说，或对抽象事理的阐释，或对常见现象的剖析使人们对事物的形态、构造、性质、种类、成因、功能、关系或对事理的概念、特点、来源、演变、异同等有所感知、认识和了解，从而获得有关的知识。把握了说明文的这些特点，译者在翻译这类文体时，就能够剔除主观理解，力避个人色彩，选用适当字句，注重逻辑联系，准确地再现原语说明文的语言信息和文体风格等表达特点。

从微观层面来说，由于英汉两种语言在词义、语法和修辞的内涵与外延以及行文与表达习惯等方面都存在差异，加之译者对原文所表达的逻辑意义、蕴涵的文化因素以及涉及的专业知识缺乏充分的认知，常常会导致理解不准确，甚至出现理解完全错误。对原文的微观理解一般包括如下几个方面：

（1）词义的理解。英语词汇，浩如烟海；确定词义，甚是不易。这主要是因为每一个词不是孤立地存在于原文，必须根据词语所处的上下文，对英语词汇一词多义的特点、搭配关系、词义的引申与褒贬，甚至词语所处原文的主题内容等诸多因素通盘考虑后，再加以确定词语的意义。

【例1】Although discovery of MGRs is in its infancy, several UNICPOLOS panelists reported that MGRs have been most commonly found near hydrothermal vents on the deep sea bed.

【译文】尽管对海洋基因资源的探索尚处于起步阶段，但联合国海洋事务和海洋法非正式磋商进程的几个专家小组成员报告说，在深海海床的热液喷口附近，海洋基因资源已经被普遍发现。

【分析】此句中画线部分"in its infancy"，本意为"处于婴幼儿期"，根据上下文，应理解成"处于起步阶段（不成熟阶段）"，本句译文取其引申意义。

【例2】Government agencies, commercial interests, conservation groups, and researchers alike are shifting their focus from traditional management tools oriented towards managing single species, to a more precautionary, ecosystem oriented approach.

【译文】政府机构、商业利益集团、环保组织和研究人员都把关注的重点从面向管理单一物种的传统管理工具转向为加大预防，并以生态系统为主导的管理方法。

【分析】本例中的"interests"是个多义词，有"兴趣""利害关系""利益""利息""感兴趣的事"等意思。这里既不是指"interest on money"，也无"hobby"之意。而是指一些（试图左右政府决策而保护自身利益的）利益集团。

所以，【例1】和【例2】说明，在翻译过程中必须根据语境或上下文来理解一个词（组）在原文中是属于一般意义还是专门意义，是概念意义还是联想意义，然后再做出恰当的选择。

（2）语法结构的理解。海洋英语文献中有很多长句与复杂句，其语法结构与层次关系错综复杂，在翻译时，必须采用语义结构分析法，厘清整句的主从结构以及语义重心和修饰成分之间的关系，唯有如此，才有可能深刻而准确地理解原句的内涵。

【例3】Today, volcanism continues to supply the atmosphere and hydrosphere with many of the same gases—think of the white plumes that steam from active volcanoes —but at a much lower level than when the early Earth was releasing so much heat.

【译文】今天，火山运动还在持续不断地向大气圈和水圈释放大量相同的气体——设想一下那些从活火山喷涌而出的白色的羽流吧——不过，与早期地球释放出的巨大热量相比，其水平要低得多。

【分析】剔除两个破折号之间的插入成分，原句是由一个表示转折关系的并列连词“but”连接的两个并列句，而在后一并列句中，不仅为了避免上下文结构重复而省略了与前一并列句相同的主谓语成分，而且还嵌入了一个由“than”引导的表示“比较”的从句。整个句子显得长而复杂，但只要对其语法关系稍加分析，厘清全句主从关系和修饰成分，即可充分理解句意。

【例 4】The Moon probably formed some 50 million years after the Earth, which is when scientists think a Mars-sized body collided with our planet.

【译文】月球可能是在地球形成之后约 5 000 万年的时候形成的，因为科学家们认为当时有一个火星大小的物体与我们地球相撞。

【分析】本例是一个由 which 引导的表示原因的非限定性定语从句，句式结构看来并不复杂。但仔细研读，便发现这个定语从句有几个异常之处：首先，which-clause 并非紧跟先行词“the Moon”之后，而被“The Moon”之后的成分分隔开了，为分隔型定语从句；其次，从句中嵌入了一个在表意上可有可无的结构“scientists think”，似使句子结构复杂化了；最后，从内在语义逻辑来分析，这个定语从句与主句存在因果关系。

（3）逻辑关系的理解。“两种语言的翻译是以逻辑关系为基础的。”当根据语法分析和语意推敲难以厘清原文时，译者就需要借助常识和逻辑推理来确定语义。逻辑规律有时可以帮助译者顺利解决紧靠分析语法关系而不能理解语义的问题。实际上，运用逻辑关系分析理解原文是对剖析语法结构不足以梳理清楚原文意义的一种延伸和补充。

【例 5】Not surprisingly, the sea's nonrenewable resources went largely unused while supplies on land were plentiful. More and more, however, land reserves are running out, and we are turning to the sea for new resources.

【译文】由于陆上供应量充足，海洋中的不可再生资源大部分尚未被利用，这一点我们并不感到奇怪。然而，越来越多的陆地（资源）储备日渐耗尽，我们开始转向海洋以获取新的资源。

【分析】从原句结构看，本例由 Not surprisingly, ...were plentiful 和 More and more, however, ...for new resources 这两个句子组成两层意思，连接前后句之间的连接词“however”使得整个句子的“语义转折”逻辑关系十分清楚。然而，上下文之间似乎存在语义上的前后矛盾：上文提到“陆上资源供应量充足”，而下文却在担忧“陆地（资源）储备日渐耗尽”，显然不符合逻辑。但根据时间逻辑推理，问题便迎刃而解：上文在讲“过去”，下文在论“现在”。上文是指过去“陆上供应量充足，海洋中的不可再生资源大部分尚未被利用”；下文是说现在“陆地（资源）储备日渐耗尽，我们正在转向海洋以获取新的资源”。这样，上下文之间的“语义转折”关系便顺理成章了。

对逻辑关系的把握往往和对关联词或关系词的理解有关，因为正是这些词在句中支配着词、词组以及从句之间的逻辑关系。如本例的前一句“Not surprisingly，the sea's nonrenewable resources went largely unused while supplies on land were plentiful”中的从属连词“while”引导的从句与主句间逻辑关系究竟是表示“时间”或“条件”，还是表示“让步”或“对比”，抑或表示“原因”。前两者似乎从逻辑上都解释不通，根据上下文判断，人们对“海洋中的不可再生资源在很大程度上未被利用”感到“毫不奇怪”，是因为“陆上供应量充足”，很显然，“while”在本句引导的是一个表示因果逻辑关系的从句，尽管这种用法在英语里很少见。

（4）文化背景知识的理解。在海洋英语文献汉译中，也常常会碰到一些跨文化交际问题，涉及英语国家的政治、经济、历史、社会、宗教以及风俗习惯等，原文的表层意义与深层含义并不一致。如果译者缺乏必要的背景知识，仅囿于表层意义的理解，势必会造成误译。

【例 6】Earth may be considered a “Goldilocks” planet：“just right” for liquid water，and thus for life，which requires it.

【译文】地球可以被看成是一个“适合（人类）居住的”星球：“正适合”液态水的存留，因而也适合需要水的生命体居住。

【分析】如果仅从表层意思去理解，不领悟“Goldilocks”蕴含的背景知识，此句前半句可能会译成“地球可以被看成是一个金凤花姑娘似的星球”，这样处理，显然过于草率，而且与下文的“正适合液态水的存留，因而也适合需要水的生命体居住”的意义前后不一致，甚至会让读者有如坠雾中，不知所云之感。这里，作者恰当地使用了美国传统童话的一个角色“Goldilocks”——金凤花姑娘，形象生动，传神达意。在童话中，金凤花姑娘只喜欢适合自己的东西，如不冷不热的粥，不软不硬的椅子，总之是“刚刚好”的东西，所以原文作者用 Goldilocks 一词来形容“刚刚好，适合的”。

（5）专业背景知识的理解。翻译中对原文的理解，不仅与原语功底有关，也需要具备一定的专业背景知识。翻译不单是语言表层结构的转换，和其他文体作品一样，海洋英语文献原句内容中亦常常涉及某些专业知识和特定表达法，其外显意义与内涵意义并不相同。如果我们忽视这一现象，或缺乏基本的专业常识，就无法准确地传达作者的原意。

【例 7】Volcanic activity beneath the earth's crust sends sulfur-rich plumes spewing up from gashes in the seafloor into the ocean, where the plumes support unique ecosystems of animals and microbes.

【译文】地壳下的火山活动将海底裂缝中喷出的富含硫黄的烟尘送入海洋，这些烟尘在海里维系着动物和微生物赖以生存的独一无二的生态系统。

【分析】如果将原文中的两处“plumes”按词典释义理解为“羽毛、羽状物”，则翻译为火山爆发能喷发出富含硫黄的“羽状物”，而这些“羽状物”还能支撑一些动物和微生物的生命，让人感到不知所云。其实，懂专业的人一看便知道，“plumes”是流体力学中的一个专业用语，意为“羽流”，即“一柱流体在另一种流体中移动”（a column of one fluid moving through another），实际上是指“烟尘或尘埃”，即活火山喷发出来的气体遇水冷却变为烟尘在水圈中移动。

【例 8】Exploratory drilling is done from drill ships or partly submerged or elevated platforms that can be towed from place to place and anchored in position on sea floor.

【译文】利用钻井船或半潜升式钻井平台进行钻探，这些半潜升式钻井平台会从一个地方被牵引到另一个地方，然后固定在海底的适当位置。

【分析】本句结构并不复杂，包含一个由 that 引导的定语从句。如果仅从语法结构关系分析，译者很有可能将此句理解为 that-clause 是用来修饰主句中的整个先行词（语）“drill ships or partly submerged or elevated platforms”，即翻译成“钻井船或半潜升式钻井平台会从一个地方被牵引到另一个地方，然后固定在海底的适当位置”。但事实上，that-clause 只修饰先行词（语）的后半部分，因为“钻井船”可在海上自行移动，而只有“半潜升式钻井平台”才需要被驳船牵引到合适的海域进行海底固定后，实施钻探作业。

2. 表达 “译无止境，妙在表达。”在翻译过程中，译者在自己对原作理解的基础上，力求忠实、准确而完整地传达原作的意旨，再现原作的语体风格。确定词（语）义重要，表达得体在一定意义上更为重要，否则，译者的努力将会变成无用之功。然而，由于译者对文本内容的理解程度不同，行文风格迥异，由同一文本产生的译文质量可能差异很大。即使译者对同一文本都有准确而深刻的理解，但其目的语言素养与功底以及翻译技巧与方法的掌握程度不一样，译文质量的优劣高低也会受到影响。因此，就英译汉而言，译者必须在理解原作的前提下，借助翻译中常用的各种技巧与方法，尽量发挥母语表达的优势，克服因两种语言结构和文化习惯差异带来的表达上的束缚，采用符合本民族语言习惯和文化特点的表达方法来再现原作的意义与风格。

(1) 借助翻译技巧（方法）表达。译文的质量最终取决于译文的表达水平，译者千万不要误认为用自己的母语表达就可驾轻就熟，从而掉以轻心，必须对照原文，反复斟酌、判断，选择恰如其分，并与原文意义、风格相符合的译文表达。这就要求译者在最大限度忠实于原作的基础上，借助自己母语的表达优势，调动自己的形象思维和逻辑思维，灵活运用各种翻译技巧和方法，包括直译法、意译法、词类转换法、增益法和省译法等，选择表达规范、语义流畅的译文。

【例 9】Discarded plastic bags, six pack rings and other forms of plastic waste which finish up in the ocean present dangers to wildlife and fisheries.

【译文】废弃的塑料袋、六环包以及其他形式的塑料垃圾会最终进入海洋，给野生动物和渔业带来危害。

【分析】本例译文与原文基本上是词字相对，天衣无缝，可采用直译法处理，既不损形，也不害义。这当然是最理想、最便捷的译法。但在多数情况下，原文的思想内容与译文的表达形式存在冲突与矛盾，如：

【例 10】The most restrictive MPAs are the marine reserves, also known as "ecological" reserves or "no-take" reserves, marine reserves are special class of MPA.

【译文】最具限制性的海洋保护区是海洋自然保护区，也称为"生态"自然保护区或"禁捕"自然保护区，自然保护区是一类特殊的海洋保护区。

【分析】本例翻译采用直译、意译、增益三种技巧：整句大体上采用了字句相对的直译，但"no take"加了引号，说明不能直译为"不能拿"，根据上下文，可理解为"禁捕"。同时，"reserves"增译为"自然保护区"，可见"自然"一词，原文形式上并不存在，却蕴含在其意义之中，这类保护区是海洋保护区中非常特殊的一类。

另一种常用来增强译文表达规范的技巧是词类转换。英汉语言各有特点，汉语中多用动词，无词性变化，动词常连用，因此，常会将其他词类转译成动词或名词。

【例 11】The use of tidal energy is pollution-free and relatively efficient, but the resulting changes in the tidal patterns can be highly destructive to the nearby environment.

【译文】利用潮汐能既无污染又相对高效，但是随之而产生的潮汐模式改变对周边环境可能会极具破坏力。

【分析】本例中"use"为名词，"destructive"为形容词，但分别译成了"动词"和"名词"。这些都是利用了词类转换技巧，使译文在忠实于原文的前提下更为贴切、通顺和地道。

（2）理顺逻辑关系表达。每一个民族都有自己独特的思维方式和表述逻辑，尤其是原文中隐而不露的一些内在的逻辑关系，必须在译文表达中准确无误地传达出来，否则，译文与原文可能会貌合神离。

【例 12】While being toxic to marine life，polycyclic aromatic hydrocarbons，the components in crude oil，are very difficult to clean up，and last for years in the sediment and marine environment.

【译文】多环芳烃——原油中的成分，不仅会毒害海洋生物，而且很难清除，还会在沉积物和海洋环境中残存多年。

【分析】本例中的“while”在词典释义和实际使用中的解释通常为“当…”“一边…，一边…”“尽管”“既然”等意思。很显然，这里“while”引导的结构与下文既没有时间上的“先后顺序”或“同时发生”的逻辑关系，也不表示“因果联系”或“转折关系”，而是存在一种递进关系。其实，在英语中，由引导定语从句的关系代词或关系副词引导的从句，形式上是一种定语从句，但在很多情况下却是一种状语化的定语从句（adverbialized attributive clause），用来表示原因、结果、目的、条件、时间和让步等逻辑关系。如：

【例 13】Most of animals and plants in the sea，which are very important to Man as a source of food，seem unfit for human consumption.

【译文】尽管海洋中的大多数动植物是人类非常重要的食物来源，但似乎不适合食用。

【分析】本例将 which-clause 译成表“让步”的状语从句，即“尽管…”，似乎更能彰显上下文之间的内在逻辑，且更符合译语的表达习惯和规范。

【例 14】Marine pollution occurs when harmful effects，or potentially harmful effects，can result from the entry into the ocean of chemicals，particles，industrial，agricultural and residential waste，or the spread of invasive organisms.

【译文】海洋污染起源于进入海洋的化学物质、微粒、工农业和生活废弃物，或入侵性生物在海洋中传播而引起的不良后果或潜在的有害影响。

【分析】本例如果按照原文的逻辑顺序来翻译“result from”（因…而导致：表示前者是后者结果），译文难免会留下“翻译痕迹”，但上述译文将主宾逻辑关系倒过来，采取“正话反说”，即后者是前者的原因（相当于 result in），更符合译语的表达习惯和规范。

（3）遵循译语规范表达。在英译汉中，应充分发挥汉语独有的表达优势。

在海洋英语文献翻译中，适当地使用一些文言文词语、四字结构，如果使用得当，不仅会使译文和原文的语言风格或专业特点更趋一致，同时也可以使译文增彩生色。

【例 15】Single species management（SSM）is a management technique that，like the name implies，focuses the management on a single species.

【译文】单一物种管理是一种管理方法，顾名思义，就是把管理的焦点集中在单一物种上。

【分析】这里，利用汉语的四字结构“顾名思义”来翻译“like the name implies”，既言简意赅，又传神达意。

另外，从语段层面来看，汉语重意合轻形合，而英语则讲形合轻意合。正如著名翻译理论家奈达（E. Nida）在其著作《译意》（*Translating Meaning*）一书中所提到的“就汉语和英语而言，也许在语言学中最重要的一个区别就是意合与形合的对比。”因此，在汉译时，绝不可受制于原文结构的束缚，应按照汉语的表达习惯与规范组织与调整行文。

【例 16】The final requirement for patent protection under the TRIPS agreement is that the invention be disclosed in a publication so as to enable a person skilled in the relevant art to reproduce the invention.

【译文】在 TRIPS 协议框架下的专利保护的最后一个要求，就是发明成果应公开出版，让相关行业的能工巧匠都能进行创造发明。

【分析】为了说明“公开出版发明成果”的目的是“让相关行业的能工巧匠都能进行创造发明”，原文必须用“so as to”（以便，为了）这一结构来引导一个表目的的状语，这种刻板的注重形合表达的形式是由英语的句法结构的特点所决定的，否则，整个句子的内在逻辑关系就不甚明朗；而汉语讲究的意义关系往往存在于此种语言使用者的感悟之中，在表达上追求语言规范和结构简约。因此，为了避免译文的洋腔洋调，使表达更合规、更地道，在翻译英语中 so...that，such...that，so that，too...to 等诸如此类的结构时，大多数情况下是无需按其释义进行汉译的。

总之，理解与表达这两个步骤是一个问题的两个方面，很难决然分开。一方面，理解过程是译者通过自身的原语知识、译语功底、文化修养和逻辑常识确定原文语义，选择得体表达的过程；另一方面，表达过程又是在进一步深化对原文的理解，不断地修正和优化译文质量。就英汉翻译来说，译文质量的优劣与高低在一定程度上取决于对两种语言的驾驭能力。只有具备扎实的英语功底，才能对原语有准确透彻的理解；良好的汉语拿捏能力有助于译者在表达中得心应手、文笔流畅；掌握一定的文化和相关专业背景知识也是理解与表达过程中必不可少的一种能力。

Questions for Discussion:

1. What is the relationship between comprehension and reproduction in translation?

2. Why is it indispensable for a translator to master the English language and the Chinese language as well in English translation?

Ⅳ. Exercise 1: Sentences in Focus　练习1：单句翻译

1. Translate the following sentences into Chinese

(1) And most importantly, why has liquid water persisted on Earth when it doesn't appear to have done so elsewhere in the solar system?

(2) The rocky planets, like Venus, Earth, and Mars, formed where it may have been too hot for water and other volatiles that form our hydrosphere and atmosphere to be present in high concentrations.

(3) Earth was too hot to sustain water on its surface, which boiled and evaporated into the blistering atmosphere.

(4) An outer crust, like the skin of an onion, composed of light elements, was extracted from the mantle as volcanic rocks.

(5) This differentiation—the separation of the planet into layers that are less dense as they extend outward from the core—happened very quickly by geological standards: within the first 100 million years after the Earth's formation.

(6) This theory holds that some mass was added to the Earth, but the impact caused most of the material (including a lot of material from Earth) to fly off and get trapped in the Earth's gravitational field.

(7) The Moon's gravitational pull on the Earth causes the Earth to expand and contract slightly, and causes the oceans to move in tidal rhythm.

(8) The outside-in model proposes that the Earth formed without water, which came with other volatiles from the meteorites or comets that bombarded the young planet.

(9) Meteorites, on the other hand, appear to be the building blocks of our planet because they've been found to have a composition similar to that of

the early Earth.

(10) The Earth's mantle is estimated to contain between three to six times as much water as in the oceans, so it's perfectly feasible that our surface water came from inside the Earth.

(11) Earth scientists believe that during the first few hundred million years after the solar system formed, while the Earth was differentiating into a core, mantle, and crust, the planet started degassing ("burping" volatiles) through volcanic activity.

(12) Today, volcanism continues to supply the atmosphere and hydrosphere with many of the same gases—think of the white plumes that steam from active volcanoes—but at a much lower level than when the early Earth was releasing so much heat.

(13) While our Earth's ocean and atmosphere evolved into a life-nurturing system, our neighboring planets fared less well.

(14) The poles of Mars are believed to contain frozen water and carbon dioxide ice caps that extend underground, and recent Rover expeditions yielded evidence that liquid water once ran on its red surface.

(15) All these things contribute to an exquisite dynamic system that brought the planet's oceans into being and protects the organisms that inhabit them.

2. Translate the following sentences into English

(1) 对于具有变性的物理学来说，世界是由分子、原子，固体、液体、气体，大气圈、水圈、岩石圈和生物圈组成的物质世界。

(2) 泛大陆大约于 2.25 亿年前由于板块构造的地质过程而发生了分离，最终将地球上的各大陆块分成了今天所看到的大陆。

(3) 尽管在某些区域会稍厚或稍薄，大陆地壳平均厚度为 25 英里（40 千米）。洋壳的厚度通常只有约 5 英里（8 千米）。

(4) "火山作用非常重要，因为它代表着行星的脉搏，" 他补充道。

(5) 一项最近对月球岩石的分析发现，它们的水含量等同于地球的上层地幔——地壳下的类熔岩岩石层。

(6) 而且，即使把蟹从海滩上带回来放在黑暗中，它们仍然保持它们的潮汐节奏。

(7) 地心引力从没有完全消失过，它只是变弱。每个有质量的物体，包括你和我，都有所谓的万有引力。

(8) 但是在此之前，地球物理学者们认为，这种俯冲带大地震仅仅在年轻的海洋地壳挤入地幔处才可能发生。

(9) 两个星期前，美国天文学家宣布发现了环绕另一颗恒星的金色行星，其大小和温度适合于外星生命生活。

(10) 科学家警告说，如果另一场地震袭击这个地区，巴东（Padang）市附近的苏门答腊（Sumatra）岛海岸将下沉数十厘米。

(11) 三天前拍摄的卫星照片显示，由于冰块的融化，著名的西北和东北通道已畅通，这使船只在北极冰帽附近海域航行成为可能。

(12) 这个岛国实际上是大西洋中脊的一个山顶，这是世界上最长的山脉之一。

(13) 1991年，位于东太平洋海隆的一处大型火山爆发，熔岩几乎横扫了那片海底，堵塞了热液喷口，造成喷口附近海洋生物的死亡。

(14) 太平洋海沟的所有地方（太平洋火环带）应该立刻疏散，因为将有地震和火山爆发。

(15) 而且，如果海洋里的水能被移走，海底、宽阔的峡谷、崎岖不平的山脉和海底河流将呈现出一种令人无法想象的景象。

Ⅴ. Exercise 2：Passages in Focus　练习2：短文翻译

1. Translate the following passages into Chinese

The ocean is not just where the land happens to be covered by water. The sea floor is geologically distinct from the continents. It is locked in a perpetual cycle of birth and destruction that shapes the ocean and controls much of the geology and geological history of the continents. Geological processes that occur beneath the waters of the sea affect not only marine life，but dry land as well. The processes that mold ocean basins occur slowly，over tens and hundreds of millions of years. On this timescale，where a human lifetime is but the blink of an eye，solid rocks flow like liquid，entire continents move across the face of the earth and mountains grow from flat plains. To understand the sea floor，we must learn to adopt the unfamiliar point of view of geological time. Geology is very important to marine biology. Habitats，or the places where organisms live，are directly shaped by geological processes. The form of coastlines；the depth of the water；whether the bottom is muddy，sandy，or rocky；and many other features of a marine habitat are determined by this geology. The geologic history of life is also called Paleontology.

In the years after World War Ⅱ, sonar allowed the first detailed surveys of large areas of the sea floor. These surveys resulted in the discovery of the mid-oceanic ridge system, a 40,000 mile continuous chain of volcanic submarine mountains and valleys that encircle the globe like the seams of a baseball. The mid-oceanic ridge system is the largest geological feature on the planet. At regular intervals the mid-ocean ridge is displaced to one side or the other by cracks in the earth's crust known as transform faults. Occasionally the submarine mountains of the ridge rise so high that they break the surface to form islands, such as Iceland and the Azores.

The portion of the mid-ocean ridge in the Atlantic, known as the Mid-Atlantic Ridge, runs right down the center of the Atlantic Ocean, closely following the curves of the opposing coastlines. The ridge forms an inverted Y in the Indian Ocean and runs up the eastern side of the Pacific. The main section of ridge in the eastern Pacific is called the East Pacific Rise. Surveys also revealed the existence of a system of deep depressions in the sea floor called trenches. Trenches are especially common in the Pacific.

It was the discovery of the magnetic anomalies on the sea floor, together with other evidence, that finally led to an understanding of plate tectonics. The earth surface is broken up into a number of plates. These plates, composed of the crust and the top parts of the mantle, make up the lithosphere. The plates are about 100 km thick. As new lithosphere is created, old lithosphere is destroyed somewhere else. Otherwise, the earth would have to constantly expand to make room for the new lithosphere. Lithosphere is destroyed at the trenches. A trench is formed when two plates collide and one plate dips below the other and slides back down into the mantle. This downwards movement of the plate into the mantle is called subduction. Because subduction occurs at the trenches, trenches are often called subduction zones. Subduction is the process that produces earthquakes and volcanoes, also underwater. The volcanoes may rise from the sea floor to create chains of volcanic islands.

We now realize that the earth's surface has undergone dramatic alterations. The continents have been carried long distances by the moving sea floor, and the ocean basins have changed in size and shape. In fact, new oceans have been born. Knowledge of the process of plate tectonics has allowed scientists to reconstruct much of the history of these changes. Scientists have discovered, for example, that the continents were once united in a single super continent called Pangaea that began to break up about 180 million years ago. The continents have since moved to their present position.

2. Translate the following passages into English

传统上，大洋分为四大洋盆。太平洋最深且最大，几乎和其他所有洋盆加在一起的面积一样大。大西洋比印度洋要大些，但这两大洋的平均深度很相似。北冰洋最小且最浅。与这几大洋盆相连或处于其边缘的是形形色色的浅海，如地中海、墨西哥湾和南中国海。

地球主要由三个层级构成：富含铁的地核，半塑型的地幔和薄薄的外壳。地壳是人们最熟悉的地球层。与较深的层级相比，它厚度极薄，就像一层坚硬的皮肤漂浮在地幔的顶部。地壳的成分与特点在海洋与大陆之间差异极大。

我们通常把大洋看成是四个独立的实体，但实际上它们是相互连接的。只要从南极看世界地图，这一点就能轻易被看出。从这个视角看，不言而喻，太平洋、大西洋和印度洋只是一个巨大海洋系统的几大分支。这几大洋盆的相互连接可以使海水、各种物质以及某些生物体从一个海洋流动到另一个海洋。由于海洋实际上是一个巨大的相互连接的系统，海洋学家谈论的常常是一个单一的世界海洋。他们也把环绕南极的那片连续不断的水域称为南大洋。

海洋与大陆之间的地质差异是由岩石本身的物理和化学差异造成的，而与岩石是否被水覆盖无关。被水所覆盖的地球的一部分，即海洋，因基岩的特性而被水覆盖了。

构成洋底的大洋地壳岩石是由统称为深色玄武岩的矿物质组成的。大多数大陆岩石隶属于称为花岗岩的一般种类。它与玄武岩的矿物成分不同，颜色普遍较淡。洋壳的密度比陆壳要大，尽管这两者比下部地幔的密度要小一些。大陆可以看成是许许多多厚厚的地壳块漂浮在地幔之上，就像冰山漂浮在水里一样。洋壳也漂浮在地幔之上，但是由于其密度较大，所以没有漂浮得像陆壳一样高。这就是为什么大陆会高出海平面并处于干涸状态，而洋壳会处于海平面之下并被水覆盖的原因。洋壳和陆壳的地质年龄也有区别。最古老的大洋岩石还不到两亿年，按地质学标准衡量，这已经是相当年轻了。另一方面，大陆岩石可能十分古老了，已有38亿年的历史。

Unit Two Marine Biological Resources

第二单元 海洋生物资源

Ⅰ. Warm-up for Theme-related Words/Expressions 词汇预习

anchoveta *n.* 南美鳀
anchovy *n.* 鳀鱼
bank *n.* 浅滩
biomass *n.* （单位面积或体积内）生物量
bulldoze *v.* 用推土机清除
bycatch *n.* 兼捕
Cape Cod 科德角
captivity *n.* 俘虏；圈养
chlorophyll *n.* 叶绿素
cod *n.* 鳕鱼
cohort *n.* 世代；年龄组
crustacean *n.* 甲壳纲动物
dredging *n.* 清淤
driftnetting *n.* 流刺网捕鱼
haddock *n.* 黑线鳕鱼
handlining *n.* 手钓
herring *n.* 鲱鱼
hibernation *n.* 冬眠；过冬
integral *n.* 构成整体所必需的
juvenile *a.* 幼年的
Lofoten *n.* 罗弗敦群岛
longlining *n.* 延绳钓
menhaden *n.* 鲱鱼
metabolism *n.* 新陈代谢
mollusk *n.* 软体动物
mortality *n.* 死亡数，死亡率
mullet *n.* 鲻鱼
ornamental *a.* 观赏性的；装饰性的
oyster *n.* 牡蛎
photic zone 透光区
phytoplankton *n.* 浮游藻类，浮游植物
protein *n.* 蛋白质
purse seine 围网
recruitment *n.* 补充量
runoff *n.* 径流
salmon *n.* 鲑鱼
scallop *n.* 扇贝
seining *n.* 围网捕捞
shellfish *n.* 贝类
squid *n.* 鱿鱼
the Georges Bank 乔治沙洲
the Grand Banks （纽芬兰）大浅滩
trawler *n.* 拖网渔船
trawling *n.* 拖网捕捞
tuna *n.* 金枪鱼
turbulence *n.* 湍流；动荡

upwelling *n.* 上涌；上升流
whopping *a.* 巨大的
yellowfin *n.* 黄鳍金枪鱼
yellowtail flounder 美洲黄盖鲽
yield-per-recruit 单位补充渔获量

Ⅱ. Lead-in & Reading 课文导读

Ocean Fishery Resources

(1) The ocean is one of Earth's most valuable natural resources. It provides food in the form of fish and shellfish—about 200 billion pounds are caught each year.

Fishing Facts

(2) The oceans have been fished for thousands of years and are an integral part of human society. Fish have been important to the world economy for all of these years, starting with the Viking trade of cod and then continuing with fisheries like those found in Lofoten, Italy, Portugal, Spain and India. Fisheries of today provide about 16% of the total world's protein with higher percentages occurring in developing nations. Fisheries are still enormously important to the economy and wellbeing of communities.

(3) The word "fisheries" refers to all of the fishing activities in the ocean, whether they are to obtain fish for the commercial fishing industry, for

recreation or to obtain ornamental fish or fish oil. Fishing activities resulting in fish not used for consumption are called industrial fisheries. Fisheries are usually designated to certain ecoregions like the salmon fishery in Alaska, the Eastern Pacific tuna fishery or the Lofoten island cod fishery. Due to the relative abundance of fish on the continental shelf, fisheries are usually marine and not freshwater.

(4) Although a world total of 86 million tons of fish were captured in 2000, China's fisheries were the most productive, capturing a whopping one third of the total. Other countries producing the most fish were Peru, Japan, the United States, Chile, Indonesia, Russia, India, Thailand, Norway and Iceland—with Peru being the most and Iceland being the least. The number of fish caught varies with the years, but appears to have leveled off at around 88 million tons per year possibly due to overfishing, economics and management practices.

(5) Fish are caught in a variety of ways, including one-man casting nets, huge trawlers, seining, driftnetting, handlining, longlining, gillnetting and diving. The most common species making up the global fisheries are herring, cod, anchovy, flounder, tuna, shrimp, mullet, squid, crab, salmon, lobster, scallops and oyster. Mollusks and crustaceans are also widely sought. The fish that are caught are not always used for food. In fact, about 40% of fish are used for other purposes such as fishmeal to feed fish grown in captivity. For example, cod is used for consumption, but is also frozen for later use. Atlantic herring is used for canning, fishmeal and fish oil. The Atlantic menhaden is used for fishmeal and fish oil and Alaska pollock is consumed, but also used for fish paste to simulate crab. The Pacific cod has recently been used as a substitute for Atlantic cod which has been overfished.

(6) The amount of fish available in the oceans is an ever-changing number due to the effects of both natural causes and human developments. It will be necessary to manage ocean fisheries in the coming years to make sure the number of fish caught never makes it to zero. A lack of fish greatly impacts the economy of communities dependent on the resource, as can be seen in Japan, eastern Canada, New England, Indonesia and Alaska. The anchovy fisheries off the coast of western South America have already collapsed and with numbers dropping violently from 20 million tons to 4 million tons—they may

never fully recover. Other collapses include the California sardine industry, the Alaskan king crab industry and the Canadian northern cod industry. In Massachusetts alone, the cod, haddock and yellowtail flounder industries collapsed, causing an economic disaster for the area.

(7) Due to the importance of fishing to the worldwide economy and the need for humans to understand human impacts on the environment, the academic division of fisheries science was developed. Fisheries science includes all aspects of marine biology, in addition to economics and management skills and information. Marine conservation issues like overfishing, sustainable fisheries and management of fisheries are also examined through fisheries science.

(8) In order for there to be plenty of fish in the years ahead, fisheries will have to develop sustainable fisheries and some will have to close. Due to the constant increase in the human population, the oceans have been overfished with a resulting decline of fish crucial to the economy and communities of the world. The control of the world's fisheries is a controversial subject, as they cannot produce enough to satisfy the demand, especially when there aren't enough fish left to breed in healthy ecosystems. Scientists are often in the role of fisheries managers and must regulate the amount of fishing in the oceans, a position not popular with those who have to make a living fishing ever decreasing populations.

(9) The two main questions facing fisheries management are:

(a) What is the carrying capacity of the ocean? How many fish are there and how many of which type of fish should be caught to make fisheries sustainable?

(b) How should fisheries resources be divided among people?

(10) Fish populate the ocean in patches instead of being spread out throughout the enormous expanse. The photic zone is only 10 – 30 m deep near the coastline, a place where phytoplanktons have enough solar energy to grow in abundance and fish have enough to eat. Most commercial fishing takes place in these coastal waters, as well as estuaries and the slope of the Continental Shelf. High nutrient contents from upwelling, runoff, the regeneration of nutrients and other ecological processes supply fish in these areas with the necessary requirements for life. The blue color of the water near the coastlines is the result of chlorophyll contained in aquatic plant life.

(11) Most fish are only found in very specific habitats. Shrimp are fished in river deltas that bring large amounts of freshwater into the ocean. The areas of highest productivity known as banks are actually where the Continental Shelf extends outward towards the ocean. These include the Georges Bank near Cape Cod, the Grand Banks near Newfoundland and Browns Bank. Areas where the ocean is very shallow also contain many fish and include the middle and southern regions of the North Sea. Coastal upwelling areas can be found off southwest Africa and off South America's western coast. In the open ocean, tuna and other mobile species like yellowfin can be found in large amounts.

(12) The question of how many fish there are in the ocean is a complicated one but can be simplified using populations of fish instead of individuals. The word "cohort"[1] refers to the year the fish was born and is used to gather population statistics. Cohorts start off as eggs with an extremely high rate of mortality, which declines as the fish gets older. Juvenile fish close to the age where they can be fished are called "recruits" . Cohort mortality is tied in with the species of fish due to variances in natural mortality. The biomass of a particular cohort is greatest when fish are rapidly growing and decreases as the fish get older and start to die.

(13) Scientists use theories and models to help determine the number and size of fish populations in the ocean. Production theory is the theory that production will be highest when the number of fish does not overwhelm the environment and there are not too few for genetic diversity of populations. The maximum sustainable yield[2] is produced when the population is of intermediate size. Yield-per-recruit theory[3] is the quest to determine the optimum age for harvesting fish. The theory of recruitment and stock allows scientists to make a guess about the optimum population size to encourage a larger population of recruits. All of the above theories must be flexible enough to allow natural fluctuations in the fish population to occur and still gather significant data; however, the theories are limited when taking into account the effect of humans on the environment and misinformation could result in overfishing of the ocean's resources.

(14) Other factors that must be taken into account are the ecological requirements of individual fish species like predation and nutrition and why fish will often migrate to different areas. Water temperatures also influence the

behavior of ecosystems，causing an increase in metabolism and predation or a sort of hibernation. Even the amount of turbulence in the water can affect predator-prey relationships，with more meetings between the two when waters are stirred up. Global warming could have a huge economic impact on the fisheries when fish stocks are forced to move to waters with more tolerable temperatures.

(15) In many countries，commercial fishing has found more temporarily economical ways of catching fish，including gill nets，purse seines，and drift nets. Although fish are trapped efficiently in one day using these fishing practices，the number of fish that are wasted this way has reached 27 million tons per year，not to mention the crucial habitats destroyed that are essential for the regeneration of fish stocks. In addition，marine mammals and birds are also caught in these nets. The wasted fish and marine life are referred to as bycatch，an unfortunate side-effect of unsustainable fishing practices that can turn the ecosystem upside-down and leave huge amounts of dead matter in the water. Other human activities like trawling and dredging of the ocean floor have bulldozed over entire underwater habitats.

Notes:

[1] cohort　世代，亦称股，是指同一时期（通常 1 年）出生或孵化的一群个体。例如，2010 世代是指 2010 年为 0 龄，2011 年为 1 龄，2012 年为 2 龄，以此类推。

[2] maximum sustainable yield（MSY）　最大可持续产量，是指在不损害资源本身的恢复能力的情况下，每年所获取的最高产量。如果把生物及其环境作为一个整体当成一项资源看待，只要开发利用适当，这项资源可以不断自我更新，持续地向人类提供所需要的产量。但是，如果在一定的时间和空间内人类取用可再生资源过量，就会破坏资源的再生能力，造成资源衰竭。

[3] yield-per-recruit（Y/R）　单位补充渔获量，是指资源群体中某一特定年龄组，平均每补充的一尾鱼一生中所能提供的产量。在平衡状态下，不同的捕捞死亡系数会带来不同的单位补充渔获量。

Ⅲ. Teaching Focus（2）: Literal Translation，Liberal Translation and Transliteration　翻译技巧（2）：直译、意译与音译

直译、意译和音译属于翻译方法的问题。人们一般把直译和意译作为一对概念，放在一起加以探讨，与其密切相关的是原文的内容和形式这一对概念，

即如果译文既忠实于原文的内容，又忠实于原文的形式，就是直译；如果译文只忠实于原文的内容，不忠实于原文的形式，就是意译。当然，还有另外一种可能，即译文忠实于原文的形式，却不忠实于原文的内容，这样的翻译就是硬译了。

关于直译和意译的争论由来已久，这方面的研究成果也有很多。早在20世纪80年代初，《外国语》曾分上、中、下三篇发表许渊冲先生的长篇论文"直译与意译"，通过大量的译例分析，对傅雷、鲁迅、曹禺、方重、柳无忌、聂华苓等翻译大家的翻译风格进行了比较，围绕内容与形式、神似与形似等概念，探讨了直译与意译的辩证关系，即"无论是直译还是意译，都要把忠实于原文的内容放在第一位，把通顺的译文形式放第二位，把忠实于原文的形式放第三位。也就是说，翻译要在忠实于原文内容的前提下，尽可能做到忠实于原文的形式；如果通顺和忠实于原文的形式之间有矛盾，那就不必拘泥于原文的形式。"

虽然直译在英语里的表达是"literal translation"，但绝不意味着直译就是完全对译出语的句子结构亦步亦趋的翻译，这样逐字逐句的翻译是死译硬译，是每一个译者都应该尽量避免的。同理，虽然意译在英语里的表达是"liberal translation"，但绝不意味着译者可以天马行空，我行我素，任意改变原文的意义或结构，因为这样的翻译就是胡译乱译，也就背离了翻译的宗旨和本质。优秀的翻译作品一定既有直译，又有意译，而优秀的译者要做到"随心所欲而不逾矩"，一方面能够充分发挥作为译者的主观能动性，另一方面又能忠实传达作者的交际意图。

那么究竟什么时候应该直译，什么时候进行意译呢？在表达同样的信息内容时，英汉语在形式上有两种可能，一种是大致对应，即相互契合，另外一种是各不相同，即相互冲突。总体来说，由于人类对主客观世界的认识具有相对的一致性，不同语言之间在表达形式上的契合性还是很大的。但是，由于不同的民族在文化、心理、历史背景和民俗民情等方面会有差异，不同语言之间在表达形式上的差异性同样不可小觑。相应地，当两种语言在表达形式上能够契合时，就可以进行直译，而当两种语言在表达形式上无法契合时，就必须要进行意译。

总之，作为翻译的两种主要手段，直译和意译并非相互排斥、相互对立，而是相互补充、相互渗透、相得益彰。此外，两者也没有优劣之分，而是要具体问题具体分析，根据不同文体的风格，灵活使用。翻译学界已经基本达成共识，即直译是基础，意译是对直译必不可少的辅助和补充。这就意味着在翻译实践中，能够直译的时候便直译，实在不能直译的时候就诉诸意译。

1. 直译

【例 1】Atlantic herring is used for canning，fishmeal and fish oil. The Atlantic menhaden is used for fishmeal and fish oil and Alaska pollock is consumed，but also used for fish paste to simulate crab.

【译文】大西洋鲱鱼被用来生产罐头、鱼粉和鱼油。大西洋油鲱被用来生产鱼粉和鱼油，阿拉斯加狭鳕被用于消费，但也被用来做鱼酱，以仿冒螃蟹的味道。

【分析】本例运用了直译的手法，因为译文和原文在形式和内容上基本是契合的，顺序也基本对应。

【例 2】Heavier elements sank to form the Earth's core，which is rich in iron and nickel. The remaining elements went into forming the rocky mantle that makes up most of the planet's volume（84%）.

【译文】比较重的元素沉下去，形成地球的核心，这里富含铁和镍。其余的元素形成岩石地幔，占据了这个行星体积的大部分（84%）。

【分析】在本例中，除了最后一部分顺序上有所调整之外，基本上也做到了形式和内容上的对应，因此也算是直译。

严格意义上的直译要求译文和原文在形式和内容上完全对应，这就意味着对于简短的句子或许还有可能直译，句子越长，结构越复杂，直译的可能性就越小。实际上，要想找到一个严格意义上的直译例句并不那么容易，而意译的例子却可以信手拈来。

2. 意译

【例 3】Fishing activities resulting in fish not used for consumption are called industrial fisheries.

【译文】如果捕捞的渔获物不用于消费，这样的捕捞作业就被称为工业渔业。

【分析】这句话意译的迹象比较明显，首先可以发现 resulting in 似乎没有被翻译出来，而实际上在“捕捞的渔获物”这个表达中已经体现了其意义，如果将其翻译成“造成”甚或是“带来”都会过于生硬。另外值得注意的是在翻译过程中句式所发生的变化：英语里一个被动语态的句子被翻译成了两个分句。

这里要注意一点，海洋英语文献是科技文体，以传递信息为主，和以传递情感为主的文学文体相比，直译的因素更多，即使是意译，也属于浅层意义上的意译，一般体现在对句子和句子成分先后顺序的调整上，而对于文学性比较强的文体，则需要深层意义上的意译，也就是说不仅要改变句子和句子成分的先后顺序，甚至要对其中的表达本身做出改变，以传达出作者想要表达的意蕴，达到作者想要实现的交流目的，这尤其体现在对于典故和隐喻的处理上。

在海洋英语文献中，绝大多数句子的翻译都是直译和意译的结合，请看下面这个例句：

【例 4】The ocean is one of Earth's most valuable natural resources. It provides food in the form of fish and shellfish—about 200 billion pounds are caught each year.

【译文】海洋是地球最宝贵的自然资源之一，它向人类提供鱼类和贝类食物，每年的捕获量大约有 2 000 亿磅。

【分析】这句话的翻译建立在直译的基础之上，例如，在翻译的过程中，主谓成分都没有发生变化。后半句则有很多意译的成分，这不仅体现在句子顺序的差异上，还体现在词性的转换上，如“about 200 billion pounds are caught each year”被译成了“每年的捕获量大约有 2 000 亿磅”，顺序几乎完全被颠倒过来，被动语态被转化成了主动语态，动词“be caught”被翻译成了名词“捕获量”。试想一下，如果逐字直译，会出现什么情况呢？“它提供食物，以鱼类和贝类的形式——大约 2 000 亿磅被捕获每年”，这样汉语译文支离破碎，佶屈聱牙，不要说在笔译中不可接受，即便在口译中也同样是不可接受的。

【例 5】Global warming could have a huge economic impact on the fisheries when fish stocks are forced to move to waters with more tolerable temperatures.

【译文】全球变暖可能会对渔业产生巨大的经济上的影响，因为鱼群会被迫迁徙到温度更加适宜的水域。

【分析】这句话的翻译更多是直译，但是也有意译的成分，尤其是在对后置限定成分的处理方面，还有就是表示时间的连词“when”被灵活翻译成了“因为”。

通过上述几个例子可以看出，直译和意译之间并无明确的分界。与其说这是两种翻译方法，不如说是两种指导理念，因为无论是直译，还是意译，都会涉及很多具体的翻译策略，如后面会讲到的词类转换、增益、省译、正译和反译、语态转换等。

3. 音译 音译即译音，也就是用一种文字符号来表达另一种文字符号的发音，因此，也可以将其视为直译的一种方式。音译的优势体现在如下两个方面：首先，可以保留原语的异国情调，丰富译入语的语言。对于像英语这样的开放性语言来说，这一优势尤其明显。正是因为英语兼收并蓄，来者不拒，大量吸收外来词，今天才成为世界上词语最丰富的语言。其次，对于具有丰富文化内涵的词语，音译可以避免意译不当所导致的文化亏损。例如，汉语中的“阴”“阳”和“风水”蕴含着丰富的文化内涵，除了音译外，任何译法都很难充分表达其意蕴。

据《春秋谷梁传》记载，孔子有“名从主人，物从中国”之说，意思是说对于人名和地名这样的专有名词，要按照所在国家或民族的语言的读音去翻译。根据刘宓庆在《文体与翻译》中的说法，和其他语言相比，汉语以音译手段吸收外来词语并不算多，但是音译法依然是一种不可忽视的翻译手段，不但像人名和地名这样的专有名词必须音译，许多在汉语中找不到准确对应的外来词语也不得不先借助音译，然后才找到合适的意译词语来替代，这方面的例子有很多，如“laser”一词最初被音译为“镭射”，后来则被改译为“激光”，还有一些现在我们现在已经习以为常的词语，如“电话”“民主”“科学”“总统”，也都曾分别有“德律风”“德谟克拉西”“赛恩斯”“柏理玺天德”这样的过渡译法。而有的音译词则被汉语所吸收，一直沿用下来，如“logic”“humor”“hysterical”被分别翻译成了“逻辑”“幽默”“歇斯底里”。

前面在讲解直译和意译的关系时说过，能够直译的不要意译，而在处理外来专有名词时，要遵循的一个原则是能意译的不要音译，只有在没有相应的意译词语时才音译。在音译时，英语发音应以国际音标为准，而汉语发音应以标准汉语拼音为准。此外，音译要遵循如下几个原则：

（1）约定俗成原则。本书中的很多人名和地名，国内早已有统一的译名，如 Virgin Islands 一般被音译成“维京群岛”或“维尔京群岛”，如果另辟蹊径，意译成“处女群岛”，反而让人不知所云。是不是在地名中出现的词语都要音译呢？也不尽然，如 Channel Islands 传统上一直被意译为“海峡群岛”，而不是音译。而大西洋中的 Canary Islands，中文译名既有音译“加纳利群岛”，又有意译“金丝雀群岛”，而前者似乎使用更加广泛。诸如此类，约定俗成，都不宜擅自更改。

（2）联想性原则。在音译时，如果能够巧妙运用原语发音和译入语意义上在联想上的巧合，将会产生出人意料的效果。如“salmon”在中文里被译成“三文鱼”，既忠实于原语的发音，又能让人联想到这种鱼橙红色肌肉上那一道道白色的脂肪条纹。这种联想性原则在商标词语的翻译中作用尤其突出，不仅可以实现其审美功能，又可以达到很好的推销效果，如“Coca-cola”“Benz”“Sprite”和“Safeguard”分别被翻译成“可口可乐”“奔驰”“雪碧”和“舒肤佳”，不但很好地译出了原语的发音，而且让人产生美好的联想，岂不妙哉？

（3）音义结合原则。其实在联想性原则中，这一点已经有所体现。根据这一原则所进行的音译，既能保留异国情调，又能表达词语的指称意义。例如，已经被我们习焉不察的“鲨鱼”其实是对 shark 的音译；“tuna”通常的译法是“金枪鱼”，但是在有些地方又被译为“吞拿鱼”；“sardine”被翻译成“沙丁鱼”；“mackerel”被翻译成“马鲛鱼”。在日常生活中，类似的例子更是俯拾皆是，不胜枚举，如“beer”“pickup”“bowling”“ballet”分别被翻译成了“啤酒”“皮卡”“保龄球”和“芭蕾舞”等。

（4）统一性原则。专有名词的翻译重在统一，对于学术研究来说，这种术语的统一意义重大，而在这方面，我们还有很多工作要做。如导致整个世界气候模式发生异常变化的“El Niño”和“La Niña”，两个词都源自西班牙语，意思分别为“圣婴”和“圣女”，但是说起“圣婴现象”和“圣女现象”，可能很多人都不知所云，因为音译的“厄尔尼诺”和“拉尼娜”已经深入人心。在术语的统一方面，现在有关机构已经出版了人名和地名的译名手册，每一个专业领域也都有专门的术语词典，作为译者应该常备案头，勤于查阅。

Questions for Discussion:

1. What is literal translation? What is liberal translation? What is the relationship between the two ways of translating?

2. What are the principles of transliteration? Please give more examples to illustrate each principle.

Ⅳ. Exercise 1: Sentences in Focus 练习 1：单句翻译

1. Translate the following sentences into Chinese

(1) The oceans have been fished for thousands of years and are an integral part of human society.

(2) Fisheries of today provide about 16% of the total world's protein with higher percentages occurring in developing nations.

(3) The word “fisheries” refers to all of the fishing activities in the ocean, whether they are to obtain fish for the commercial fishing industry, for recreation or to obtain ornamental fish or fish oil.

(4) Due to the relative abundance of fish on the continental shelf, fisheries are usually marine and not freshwater.

(5) The number of fish caught varies with the years, but appears to have leveled off at around 88 million tons per year possibly due to overfishing,

economics and management practices.

(6) Fish are caught in a variety of ways, including one-man casting nets, huge trawlers, seining, driftnetting, handlining, longlining, gillnetting and diving.

(7) The fish that are caught are not always used for food. In fact, about 40% of fish are used for other purposes such as fishmeal to feed fish grown in captivity.

(8) The amount of fish available in the oceans is an ever-changing number due to the effects of both natural causes and human developments.

(9) Due to the constant increase in the human population, the oceans have been overfished with a resulting decline of fish crucial to the economy and communities of the world.

(10) Scientists are often in the role of fisheries managers and must regulate the amount of fishing in the oceans, a position not popular with those who have to make a living fishing ever decreasing populations.

(11) The question of how many fish there are in the ocean is a complicated one but can be simplified using populations of fish instead of individuals.

(12) Cohort mortality is tied in with the species of fish due to variances in natural mortality.

(13) Production theory is the theory that production will be highest when the number of fish does not overwhelm the environment and there are not too few for genetic diversity of populations.

(14) Even the amount of turbulence in the water can affect predator-prey relationships, with more meetings between the two when waters are stirred up.

(15) The wasted fish and marine life are referred to as bycatch, an unfortunate side-effect of unsustainable fishing practices that can turn the ecosystem upside-down and leave huge amounts of dead matter in the water.

2. Translate the following sentences into English

(1) 生物量取决于绿色植物和其他生产者所固定碳的量。

(2) 人工饲养的大猩猩表现出了显著的聪明才智，甚至学会了简单的人类手语。

(3) 大型流刺网的广泛使用是一种破坏性的捕捞方式，对世界海洋里的海洋生物资源构成威胁。

（4）河口区域是淡水河流和咸水海洋交汇的地方，是钓鱼、划船、观鸟和徒步旅行的好去处。

（5）延绳钓可能会导致很多问题，如在捕捞某些经济鱼类的过程中杀死很多其他的海洋动物。

（6）被捕到之后扔掉或者是从渔具逃脱的鱼类死亡率是影响目前海洋渔业管理的一个重要问题。

（7）只要有捕捞，就会有兼捕，即对非目标物种的偶然捕获，如海豚、海龟和海鸟。

（8）观赏性动植物如果被偶然或故意释放到野外，就会形成繁殖群体，常会对本地生态造成灾难性的影响。

（9）在生态系统中，掠食行为是指掠食者以被掠食者为食的生物互动。

（10）中西太平洋是世界金枪鱼围网渔业的主要渔场之一，主要渔获鱼种为鲣鱼、黄鳍金枪鱼与大眼金枪鱼。

（11）幼年动物有时看起来和成年动物大相径庭，尤其是颜色。

（12）水温即使有微小的波动，也会对鱼类产生影响。

（13）休眠常常和低体温联系在一起，其作用是在没有足够食物时节省能量。

（14）如果你很难吃胖或长肌肉，这通常是新陈代谢快的第一个信号。

（15）大海里通常到处都有上升流，由风力和地形驱动，将海洋深处的养分带到表面，为海洋可持续发展提供资源。

Ⅴ. Exercise 2：Passages in Focus 练习 2：短文翻译

1. Translate the following passages into Chinese

Global fish production has grown steadily in the last five decades，with food fish supply increasing at an average annual rate of 3.2 percent，outpacing world population growth at 1.6 percent. World per capita fish apparent consumption increased from an average of 9.9 kg in the 1960s to 19.2 kg in 2012. This impressive development has been driven by a combination of population growth，rising incomes and urbanization，and facilitated by the strong expansion of fish production and more efficient distribution channels. China has been responsible for most of the growth in fish availability，owing to the dramatic expansion in its fish production，particularly from aquaculture. Its per capita apparent fish consumption also increased an average annual rate of 6.0 percent in the period 1990 – 2010 to about 35.1 kg in 2010. Annual per capita fish supply in the rest of the world was about 15.4 kg in 2010 (11.4 kg in the 1960s and 13.5 kg in the 1990s).

Despite the surge in annual per capita apparent fish consumption in developing regions (from 5.2 kg in 1961 to 17.8 kg in 2010) and low-income food-deficit countries (LIFDCs) (from 4.9 to 10.9 kg), developed regions still have higher levels of consumption, although the gap is narrowing. A sizeable and growing share of fish consumed in developed countries consists of imports, owing to steady demand and declining domestic fishery production. In developing countries, fish consumption tends to be based on locally and seasonally available products, with supply driving the fish chain. However, fuelled by rising domestic income and wealth, consumers in emerging economies are experiencing a diversification of the types of fish available owing to an increase in fishery imports.

A portion of 150 g of fish can provide about 50 – 60 percent of an adult's daily protein requirements. In 2010, fish accounted for 16.7 percent of the global population's intake of animal protein and 6.5 percent of all protein consumed. Moreover, fish provided more than 2.9 billion people with almost 20 percent of their intake of animal protein, and 4.3 billion people with about 15 percent of such protein. Fish proteins can represent a crucial nutritional component in some densely populated countries where total protein intake levels may be low.

Global capture fishery production of 93.7 million tonnes in 2011 was the second highest ever (93.8 million tonnes in 1996). Moreover, excluding anchoveta catches, 2012 showed a new maximum production (86.6 million tonnes). Nevertheless, such figures represent a continuation of the generally stable situation reported previously. Global fishery production in marine waters was 82.6 million tonnes in 2011 and 79.7 million tonnes in 2012. In these years, 18 countries (11 in Asia) caught more than an average of one million tonnes per year, accounting for more than 76 percent of global marine catches.

The Northwest and Western Central Pacific are the areas with highest and still-growing catches. Production in the Southeast Pacific is always strongly influenced by climatic variations. In the Northeast Pacific, the total catch in 2012 was the same as in 2003. The long-standing growth in catch in the Indian Ocean continued in 2012. After three years (2007 - 2009) when piracy negatively affected fishing in the Western Indian Ocean, tuna catches have recovered. The Northern Atlantic areas and the Mediterranean and Black Sea again showed shrinking catches for 2011 and 2012. Catches in the Southwest and Southeast Atlantic have recently been recovering.

2. Translate the following passage into English

世界水产养殖的产量持续增长，虽然增长速度正在放慢。根据联合国粮农组织在全球范围内收集的最新统计数据，2012 年，世界水产养殖的产量达到了有史以来的又一个新高，多达 9 040 万吨（活体重量），产值高达 1 444 亿美元，其中包括 6 660 万吨食用鱼（产值为 1 377 亿美元）和 2 380 万吨水生藻类（大部分为海草，产值为 64 亿美元）。

此外，根据有些国家的集体报告，还有 22 400 吨的非食用产品（产值为 2.224 亿美元），如用于观赏和装饰的珍珠和贝壳。在这篇分析文章中，“食用鱼”这个术语包括被用来供人类食用的有鳍鱼类、甲壳动物、软体动物、两栖动物、淡水龟和其他水生动物（如海参、海胆、海鞘和食用水母）。到本文写作时，有些国家（包括渔业大国如中国和菲律宾）已经公布了它们 2013 年暂时或最终的官方水产养殖数据。根据最新信息，联合国联农组织估计 2013 年世界食用鱼养殖业的产量为 7 050 万吨，增长了 5.8%。据估计，其中养殖的水生植物（大部分是海藻）的产量为 2 610 万吨。2013 年，仅中国就生产了 4 350万吨的食用鱼和 1 350 万吨的水生藻类。

全球水产养殖的总价可能被高估了，其原因包括一些国家汇报的是零售价、产品价或出口价，而不是初次销售的价格。虽然如此，如果按照总水平来考虑，这些数据依然有用，可以从中看出发展趋势，还可以用来比较不同种类的水产养殖和不同的养殖水生物种群在经济效益上的相对重要性。

在鱼类供应总量方面，全球水产业发展趋势变得日益重要，这种趋势一直在持续。2012 年，捕捞渔业（包括非食用目的）和水产业的鱼类总产量为 1.58 亿吨，而食用鱼养殖贡献了其中的 42.2%，这是破纪录的，因为在 1990 年和 2000 年，这个比例分别是 13.4%和 25.7%。在整个亚洲范围内，从 2008 年以来，养殖鱼类的产量就一直高于野生捕获量。2012 年，其水产养殖业的产量占到了总产量的 54%，而在欧洲，这个比例为 18%，在其他几个大洲不到 15%。

由于大部分生产国对食用鱼的需求日益增长，水产养殖产量的总体增长趋势依然相对强劲。但是，最近几年，一些主要生产国的水产养殖业产量下降，尤其是美国、西班牙、法国、意大利、日本和韩国。在所有这些国家，有鳍鱼类的产量都有下降，而其中有些国家软体动物的产量也下降了。这种下降主要是因为可以从其他生产成本相对较低的国家进口鱼类。上述国家由此产生的鱼类供应缺口是推动其他国家鼓励扩大生产的原因之一，重点是面向出口的鱼种。

从2000年至2012年，世界食用鱼养殖的产量每年平均增长6.2%，与1980年至1990年之间（10.8%）和1990年至2000年之间（9.5%）相比，都有所下降。从1980年至2012年，世界水产养殖的产量平均年增长8.6%。从2000年到2012年，世界食用鱼养殖的产量翻了一番还要多，从3 240万吨增加到了6 660万吨。

Unit Three　Marine Mineral Resources

第三单元　海洋矿产资源

Ⅰ. Warm-up for Theme-related Words/Expressions　词汇预习

accretion　*n.* 添加，冲积层
array　*n.* 排列，大批
attendant　*a.* 伴随的
benthic　*a.* 海底的，底栖的
cobalt crust　富钴结壳
configuration　*n.* 配置，结构，外形
crystalline　*a.* 结晶状的，透明的
deposit　*n.* 矿床，沉积物　*v.* 沉淀
elevation　*n.* 高地，隆起
endemic　*a.* 地方性的
erosion　*n.* 侵蚀，腐蚀
euphoria　*n.* 欢快，兴高采烈
extraterritorial　*a.* 疆界之外的
extrude　*v.* 挤出，压出
ferromanganese　*n.* 锰铁
flank　*n.* 侧面，侧翼
manganese　*n.* 锰
fracture zone　断裂带
gas hydrate　气体水合物
gravel　*n.* 碎石；沙砾
hydrothermal　*a.* 热液的
hydroxide　*n.* 氢氧化物
magma　*n.* 岩浆
nodule　*n.* 结瘤，结核
nucleus　*n.* 核，原子核
onshore　*a.* 陆上的，朝着岸上的
ore　*n.* 矿砂，矿石
platinum　*n.* 铂
polymetallic　*a.* 多金属的
pore waters　孔隙水
precipitate　*n.* 沉淀物　*v.* 沉淀
prospect　*n.* 勘探　*v.* 前景，预期
repository　*n.* 贮藏室，仓库
resurgence　*n.* 复活，再现
sediment　*n.* 沉淀物
seep　*v.* 渗漏
signatory　*n.* 签约国，签约人
silicate　*n.* 硅酸盐
solution　*n.* 溶液
submarine　*a.* 水下的，海底的
substrate　*n.* 基质，底层
sulphide　*n.* 硫化物
taxonomic classification　系统分类
tectonic　*n.* 构造的
tellurium　*n.* 碲
titanium　*n.* 钛
trace　*n.* 痕迹，微量
trove　*n.* 发现物，收藏物

Ⅱ. Lead-in & Reading 课文导读

The Sea Floor—Humankind's Resource Repository

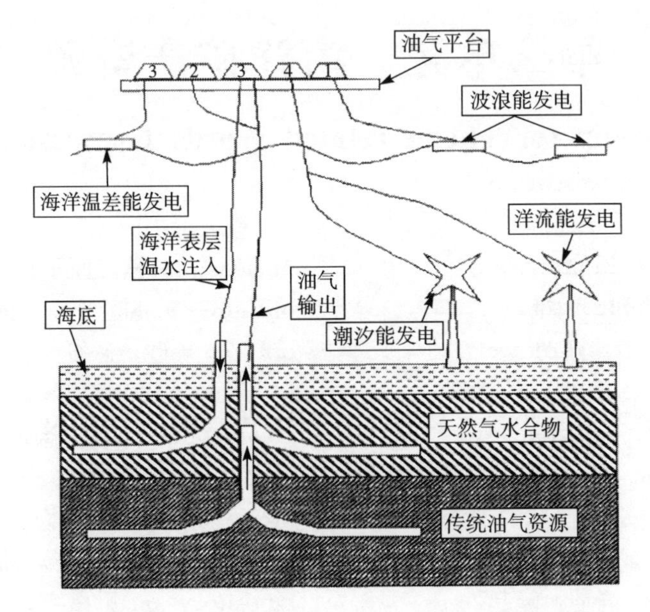

(1) The oceans hold a veritable treasure trove of valuable resources. Sand and gravel, oil and gas have been extracted from the sea for many years. In addition, minerals transported by erosion from the continents to the coastal areas are mined from the shallow shelf and beach areas. These include diamonds off the coasts of South Africa and Namibia as well as deposits of tin, titanium and gold along the shores of Africa, Asia and South America. Efforts to expand ocean mining into deep-sea waters have recently begun. The major focus is on manganese nodules, which are usually located at depths below 4,000 metres, gas hydrates (located between 350 and 5,000 metres), and cobalt crusts along the flanks of undersea mountain ranges (between 1,000 and 3,000 metres), as well as massive sulphides and the sulphide muds that form in areas of volcanic activity near the plate boundaries, at depths of 500 to 4,000 metres.

(2) Back in the early 1980s there was great commercial interest in manganese nodules and cobalt crusts. This initial euphoria over marine mining

led to the International Seabed Authority (ISA)[1] being established in Jamaica, and the United Nations Convention on the Law of the Sea (UNCLOS) being signed in 1982. Since entering into force in 1994, this major convention has formed the basis for signatories' legal rights to use the marine resources on the sea floor outside national territorial waters[2].

(3) After that, however, the industrial countries lost interest. For one thing, prices dropped—making it no longer profitable to retrieve the accretions from the deep sea and utilize the metals they contained. Also, new onshore deposits were discovered, which were cheaper to exploit. The present resurgence of interest is due to the sharp increase in resource prices and attendant rise in profitability of the exploration business, and in particular to strong economic growth in countries like China and India which purchase large quantities of metal on world markets. The industrial and emerging countries' geopolitical interests in safeguarding their supplies of resources also play a role. In light of the increasing demand for resources, those countries which have no reserves of their own are seeking to assert extraterritorial claims in the oceans.

Manganese Nodules

(4) Covering huge areas of the deep sea with masses of up to 75 kilograms per square metre, manganese nodules are lumps of minerals ranging in size from a potato to a head of lettuce. They are composed mainly of manganese, iron, silicates and hydroxides, and they grow around a crystalline nucleus at a rate of only about 1 to 3 millimetres per million years. The chemical elements are precipitated from seawater or originate in the pore waters of the underlying sediments. The elements of economic interest, including cobalt, copper and nickel, are present in lower concentrations and make up a total of around 3.0 percent by weight. In addition, there are traces of other significant elements such as platinum or tellurium that are important in industry for various high-tech products. The greatest densities of nodules occur off the west coast of Mexico (in the Clarion-Clipperton Zone), in the Peru Basin, and the Indian Ocean. In the Clarion-Clipperton Zone the manganese nodules lie on the deep-sea sediments covering an area of at least 9 million square kilometres—an area the size of Europe. Their concentration in this area can probably be attributed

to an increased input of manganese-rich minerals through the sediments released from the interior of the Earth at the East Pacific Rise by hydrothermal activity—that is, released from within the Earth by warm-water seeps on the sea floor and distributed over a large area by deep ocean currents.

(5) The actual mining process does not present any major technological problems because the nodules can be collected fairly easily from the surface of the sea floor. Excavation tests as early as 1978 were successful in transporting manganese nodules up to the sea surface. But before large-scale mining of the nodules can be carried out there are still questions that need to be answered. For one, neither the density of nodule occurrence nor the variability of the metal content is accurately known. In addition, recent investigations show that the deep seabed is not as flat as it was thought to be 30 years ago. The presence of numerous volcanic elevations limits the size of the areas that can be mined. Furthermore, the excavation of manganese nodules would considerably disturb parts of the seabed. The projected impact would affect about 120 square kilometres of ocean floor per year. Huge amounts of sediment, water, and countless organisms would be dug up with the nodules, and the destruction of the deep-sea habitat would be substantial. It is not yet known how, or even whether, repopulation of the excavated areas would occur. Since 2001 several permits have been issued to governmental institutions by the ISA to survey manganese fields. These are not for actual mining but for a detailed initial investigation of the potential mining areas. In 2006 Germany also secured the rights to a 150,000 square kilometre area—twice the size of Bavaria—for a period of 15 years. Last year, for the first time, industrial companies also submitted applications for the exploration of manganese nodule fields in the open sea in cooperation with developing countries.

Cobalt Crusts

(6) Cobalt crusts form at depths of 1,000 to 3,000 metres on the flanks of submarine volcanoes, and therefore usually occur in regions with high volcanic activity such as the territorial waters around the island states of the South Pacific. The crusts accumulate when manganese, iron and a wide array of trace metals dissolved in the water (cobalt, copper, nickel, and platinum) are deposited on the volcanic substrates. Their growth rates are comparable to

those of manganese nodules. The cobalt crusts also contain relatively small amounts of the economically important resources. Literally tonnes of raw material have to be excavated in order to obtain significant amounts of the metals. However, the content of cobalt and platinum is somewhat higher than in manganese nodules.

(7) Technologically, the mining of cobalt crusts is much more complex than manganese nodules. For one, it is critical that only the crust is removed, and not the underlying volcanic rocks. In addition, the slopes of the volcanoes are very ragged and steep, which makes the use of excavation equipment more difficult. It is therefore not surprising that cobalt crust mining is only at the conceptual stage at present. Cobalt crust mining would also have a significant impact on the benthic organisms. It is therefore vital that prior environmental impact studies are carried out. In most cases monitoring by the ISA is not possible because many cobalt occurrences are located within the territorial waters of various countries.

Massive Sulphides

(8) The third resource under discussion is a sulphur-rich ore that originates at "black smokers"[3]. These occurrences of massive sulphides form at submarine plate boundaries, where an exchange of heat and elements occurs between rocks in the Earth's crust and the ocean due to the interaction of volcanic activity with seawater. Cold seawater penetrates through cracks in the sea floor down to depths of several kilometres. Near heat sources such as magma chambers, the seawater is heated to temperatures exceeding 400 degrees Celsius. Upon warming, the water rises rapidly again and is extruded back into the sea. These hydrothermal solutions transport metals dissolved from the rocks and magma, which are then deposited on the sea floor and accumulate in layers. This is how the massive sulphides and the characteristic chimneys ("black smokers") are produced.

(9) These were first discovered in 1978 at the East Pacific Rise. For a long time it was thought that massive sulphides with mining potential were only formed on mid-ocean ridges, because the volcanic activity and heat production here are especially intense. But since then more than 200 occurrences worldwide have been identified. Experts even estimate that 500 to

1,000 large occurrences may exist on the sea floor. But there are also great differences in size. Most occurrences are only a few metres in diameter and the amount of material present is negligible. So far only a few massive sulphide occurrences which are of economic interest due to their size and composition are known. While the black smokers along the East Pacific Rise and in the central Atlantic produce sulphides comprising predominantly iron-rich sulphur compounds—which are not worth considering for deep-sea mining—the occurrences in the southwest Pacific contain greater amounts of copper, zinc and gold. They are also located in comparatively shallow water (less than 2,000 metres) and lie within the exclusive economic zones of nations near them, which makes the possible mining more technologically and politically feasible. This is because a country can decide for itself with respect to the mining of marine resources within its own exclusive economic zone. The deep sea floor outside these sovereignty limits, however, is overseen by the ISA.

(10) Present mining scenarios primarily envision the exploitation of cooled, inactive massive sulphide occurrences that are only sparsely populated by living organisms. Active black smokers are rejected for the time being because most of them contain only comparatively minor amounts of resources. Furthermore, because of the nutrient-rich waters rising from below, they provide an important habitat for numerous, and in part, endemic organisms. The largest known sulphide occurrence is located in the Red Sea, where tectonic forces are pulling Africa and the Arabian Peninsula apart. Here, the sulphides are not associated with black smokers, but appear in the form of iron-rich ore muds with high contents of copper, zinc and gold. This occurrence, at a water depth of about 2,000 metres, was discovered in the 1960s. Because of its muddy consistency, it appears that these deposits will not prove problematic to mine, and this was successfully tested in the 1980s.

(11) Of the three sea floor resources discussed here, massive sulphides are the least abundant in terms of total volume, but they are of particular interest because of their high resource content. Some mining companies have already obtained exploration licenses in national waters, and are advancing the technology for prospecting and extraction. In May 2010 the ISA even has granted one exploration license in the Indian Ocean to China. So far only

permits for research have been granted for the deep sea. In the near future the mining of copper and gold from massive sulphides is likely to commence off the coasts of Papua New Guinea and New Zealand. Mining operations had been planned to start this year, but due to the present economic recession, major metal and mining companies have experienced a decline in turnover in spite of the relatively high prices of gold, and the projects were postponed at short notice. But a recovery of the metal market is expected for the future. The companies will therefore soon be able to proceed with their plans.

The Future of Marine Mining

(12) Of the three resource types waiting to be extracted from the deep sea, the mining of massive sulphides in the exclusive economic zones (200 nautical miles) of west Pacific nations (Papua New Guinea) seems to be most feasible at present. Despite the latest economic crisis, production could start in the next few years. Because of their relatively high content of valuable metals, the mining of massive sulphides may be profitable for some companies. But the metal content of the global massive sulphides is lower than that of the ore deposits on land. It is therefore unlikely that the marine mining of massive sulphides will have a significant impact on the global resource supply.

(13) Manganese nodules and cobalt crusts present quite different prospects. The amounts of copper, cobalt and nickel they contain could without doubt rival the occurrences on land. In fact, the total cobalt is significantly more than in all the known deposits on land. About 70,000 tonnes of cobalt are presently mined on land each year and the worldwide supply is estimated at about 15 million tonnes. By comparison, a total of about 1,000 million tonnes of cobalt is estimated to be contained in the marine manganese nodules and cobalt crusts. In spite of these immense resources, sea floor mining will only be able to compete with the substantial deposits presently available on land if there is sufficient demand and metal prices are correspondingly high. Furthermore, the excavation technology has yet to be developed. The serious technological difficulties in separating the crusts from the substrate, combined with the problems presented by the uneven sea floor surface, further reduce the economic potential of the cobalt crusts for the

present. Therefore, it seems that marine mining of cobalt crusts should not be anticipated any time soon.

Notes:

[1] International Seabed Authority (ISA) 国际海底管理局，是管理国际海底区域及其资源的权威组织。根据联合国第三次海洋法会议的决议，1983 年 3 月成立了联合国国际海底管理局和国际海洋法法庭筹备委员会，简称海底筹委会。主要任务是为筹备建立国际海底管理局和国际海洋法法庭而制定有关规则、规章，处理先驱投资者申请登记问题。1994 年 11 月 16 日，《联合国海洋法公约》生效，同年，国际海底管理局在牙买加首都金斯敦宣告正式成立。

[2] territorial waters 领海，沿海国主权管辖下与其海岸或内水相邻的一定宽度的海域，是国家领土的组成部分。领海的上空、海床和底土，均属沿海国主权管辖。

[3] black smokers 海底黑烟柱，一般指海底热泉，属深海热液喷口的一种。在海底发生扩张运动的地方，沿着地壳裂口会形成热液喷口。有些喷口被称为“海底黑烟柱”，因为喷出的水柱因含有化学物质而呈现出灰色和黑色。其实，这种热液喷口的原理和火山喷泉类似，喷出来的热水就像烟囱一样。发现的热泉不仅仅有黑烟囱、还有白烟囱和黄烟囱，同时，在“烟囱林”中有各种生物生存。

Ⅲ. Teaching Focus (3): Conversion of Parts of Speech 翻译技巧（3）：词类转换

在前面的讲解中已经提到过，英汉两种语言隶属于完全不同的语系，在词汇、结构和表达等方面都有很多差异。汉语中一个词类可以充当多种句子成分，而在英语中，一个词类只能充当有限的句子成分，这一点在动词上体现得尤其明显。在英语中，动词只能充当谓语，而在汉语中，动词除了充当谓语之外，还可以充当其他多种成分，包括主语、宾语、定语和表语。因此，在英汉翻译的过程中，不能按照原文的词类“对号入座”，逐字硬译，而是要在忠实于原文意义的前提下，根据汉语的表达习惯，将原文中某些词语的词类或成分转换成汉语中的其他词类或成分，只有这样，译文才能通顺自然。这样的翻译方法就叫词类转换法。

由于英汉语之间的上述差异，在英汉翻译的过程中，词类转换的形式复杂多样。在涉海类英语翻译的过程中，转换法是使用最为广泛的方法之一，也是呈现形式最多的一种方法。概括而言，主要有以下几种形式：

1. 转译为汉语动词 汉语是一种动态的语言，动词用得多，可以几个动词或动词性结构连用，而名词用得少，而英语是一种静态的语言，其行文动词用得少，句子中往往只有一个谓语动词，大量本应由动词表达的概念是用名词来表达的。因此，在英译汉时，英语中的很多意思要用汉语的动词来表达。

（1）英语名词转译为汉语动词。

①一些习语中的主体名词可以转换成动词。

【例 1】Industrial companies also submitted applications for the exploration of manganese nodule fields in the open sea in cooperation with developing countries.

【译文】产业公司也递交申请，要和发展中国家合作，共同勘探公海中的锰结核。

【例 2】The third resource under discussion is a sulphur-rich ore that originates at "black smokers".

【译文】讨论所涉及的第三种资源是源自"黑烟柱"的硫化物矿石。

【分析】在【例 1】和【例 2】中，介词短语"in cooperation with"和"under discussion"分别被翻译成了其主体名词的动词意义，即"合作"和"讨论"。

②含有动作意味的名词可以转换成动词。

【例 3】Efforts to expand ocean mining into deep-sea waters have recently begun. The major focus is on manganese nodules, which are usually located at depths below 4,000 metres.

【译文】将海洋采矿延伸到深海水域的努力刚刚开始，主要聚焦于通常位于 4 000 米深度以下的锰结核。

【分析】"focus"在本例中本来是名词，作为主语，在汉语中被译为动词"聚焦于"，而被省略掉的主语也就相应地变成了上半句中的"努力"。

③由动词派生出来的名词可以转换成动词。

【例 4】The excavation of massive sulphides and manganese nodules is expected to begin within the next few years.

【译文】根据预期，今后几年内将开始开采大型硫化矿和锰结核。

【分析】在本例中，"excavation"被翻译成了其动词意义。当然，也可以不进行转换，译文同样也可以很通顺：根据预期，对大型硫化矿和锰结核的开采将在今后几年内开始。

（2）英语介词转译为汉语动词。英语中含有动作意味的介词可以转译为汉语的动词。

【例 5】In light of the increasing demand for resources, those countries which have no reserves of their own are seeking to assert extraterritorial claims in the oceans.

【译文】考虑到对资源的需求日益增长，那些没有属于自己的资源储备的国家正试图提出要求占有疆界之外海洋里的资源。

【分析】在本例中，介词短语“in light of”意为“鉴于”“根据”，这里可以翻译为“考虑到”。

（3）英语形容词转译为汉语动词。英语中表示知觉、情感、欲望等心理状态的形容词在系动词后面做表语时，常转译为相应的动词，如 afraid，thankful，certain，able，doubtful，grateful，cautious，sure，confident 等，这种转换相对比较简单。

【例 6】The companies will therefore soon be able to proceed with their plans.

【译文】因此，公司将很快就能够执行他们的计划。

2. 转译为汉语名词

（1）英语动词转译为汉语名词。英语中由名词派生出来的动词或由名词转用的动词，为符合汉语表达习惯，可将其还原成名词来译。

【例 7】Electrical generation from tidal power plants is characterized by periods of maximum generation every twelve hours，with no electricity generation at the sixth hour mark in between.

【译文】潮汐发电的特征是每 12 个小时出现一段发电高峰时期，而在其间的第 6 个小时不发电。

【例 8】Island tourism development is widely considered to be typified by，amongst other factors，small geographical size，distance and isolation from metropolitan centres，a limited economic base，and a lack of resources.

【译文】人们广泛认为岛屿旅游业的发展的特征包括地理空间小、远离市中心、经济基础薄弱、资源短缺等诸多因素。

【分析】在【例 7】和【例 8】中，被动语态的动词“characterize”和“typify”都被译为名词“特征”。

（2）英语形容词转译为汉语名词。英语中有些形容词加上定冠词表示某一类人，汉译为名词，这种情况也比较简单，此处不再举例。

3. 转译为汉语形容词

（1）将英语动词译为汉语名词时，修饰该动词的副词要转译成汉语形容词。

【例 9】For one，neither the density of nodule occurrence nor the variability of the metal content is accurately known.

【译文】一方面，无论是对于结核形成的密度，还是对于金属含量的差异，我们都缺乏精确的了解。

【分析】在本例中，被动语态的动词“know”被翻译成了名词“了解”，相应地，副词“accurately”被译为形容词“精确的”。再例如：

【例 10】The excavation of manganese nodules would considerably disturb parts of the seabed.

【译文】锰结核的开采会对部分海底造成很大的破坏。

（2）英语副词转译为汉语形容词。有些副词是由形容词派生出来的，可转译为形容词。

【例 11】The cobalt crusts also contain relatively small amounts of the economically important resources.

【译文】钴结壳还包含相对少量具有重要经济价值的资源。

【分析】在本例中，原文中的副词“economically”被翻译成了形容词。

4. 转译为汉语副词 英语形容词转译为汉语副词。将英语名词译为汉语动词时，修饰该名词的形容词要转译为汉语副词。

【例 12】Excavation tests as early as 1978 were successful in transporting manganese nodules up to the sea surface.

【译文】早在 1978 年，开采测试成功地将锰结核运送到海面之上。

【例 13】The ores and mineral deposits on the sea floor have attracted little interest.

【译文】海底的矿石和矿藏很少引起人们的兴趣。

【分析】在【例 12】和【例 13】中，形容词“successful”和“little”均被译为副词。

虽然这里罗列了很多种不同的转换方式，但远不能穷尽所有的可能，因为翻译中词类的转换是非常灵活的，一个句子可能有多种表达方式来表达同样的意思。因此，在翻译时必须要考虑两个关键因素：首先是怎样才能让译文更符合原文作者的意图和语气；其次是怎样才能让译文的表达符合译入语的表达习惯。

词类转换是意译的主要形式和手段，正如在第二单元关于直译和意译的讲解中所说的那样，能够直译的时候不要意译，同理，能够不转换的时候，尽量不要转换。毕竟，词性和句子成分的选择常常是作者要实现一定修辞和表达目的的手段，在此情况之下，只有忠实于其形式，才能忠实于其内容。换句话说，英汉翻译中词类转换本身不是目的，而是实现翻译通达顺畅的一种手段，完全没有必要为了转换而转换。事实上，就像讲解中所提到的那样，有些例句在不进行词类转换的情况下同样也可以翻译得很通顺。

Questions for Discussion:

1. Why conversion is necessary in the English to Chinese translation of marine English?

2. What are the most frequently used ways of conversion in English to Chinese translation? Give some examples.

Ⅳ. Exercise 1: Sentences in Focus　练习 1：单句翻译

1. Translate the following sentences into Chinese

(1) Since entering into force in 1994, this major convention has formed the basis for signatories' legal rights to use the marine resources on the sea floor outside national territorial waters.

(2) The present resurgence of interest is due to the sharp increase in resource prices and attendant rise in profitability of the exploration business, and in particular to strong economic growth in countries like China and India which purchase large quantities of metal on world markets.

(3) Covering huge areas of the deep sea with masses of up to 75 kilograms per square metre, manganese nodules are lumps of minerals ranging in size from a potato to a head of lettuce.

(4) The chemical elements are precipitated from seawater or originate in the pore waters of the underlying sediments.

(5) In addition, there are traces of other significant elements such as platinum or tellurium that are important in industry for various high-tech products.

(6) The actual mining process does not present any major technological problems because the nodules can be collected fairly easily from the surface of the sea floor.

(7) Huge amounts of sediment, water, and countless organisms would be dug up with the nodules, and the destruction of the deep-sea habitat would be substantial.

(8) Cobalt crust mining would also have a significant impact on the

benthic organisms. It is therefore vital that prior environmental impact studies are carried out.

(9) These occurrences of massive sulphides form at submarine plate boundaries, where an exchange of heat and elements occurs between rocks in the Earth's crust and the ocean due to the interaction of volcanic activity with seawater.

(10) For a long time it was thought that massive sulphides with mining potential were only formed on mid-ocean ridges, because the volcanic activity and heat production here are especially intense.

(11) While the black smokers along the East Pacific Rise and in the central Atlantic produce sulphides comprising predominantly iron-rich sulphur compounds—which are not worth considering for deep-sea mining—the occurrences in the southwest Pacific contain greater amounts of copper, zinc and gold.

(12) Present mining scenarios primarily envision the exploitation of cooled, inactive massive sulphide occurrences that are only sparsely populated by living organisms.

(13) Mining operations had been planned to start this year, but due to the present economic recession, major metal and mining companies have experienced a decline in turnover in spite of the relatively high prices of gold, and the projects were postponed at short notice.

(14) In spite of these immense resources, sea floor mining will only be able to compete with the substantial deposits presently available on land if there is sufficient demand and metal prices are correspondingly high.

(15) The serious technological difficulties in separating the crusts from the substrate, combined with the problems presented by the uneven sea floor surface, further reduce the economic potential of the cobalt crusts for the present.

2. Translate the following sentences into English

(1) 最近的研究表明巨型恒星是可以通过积累形成的，但是研究者还没有观察到这一现象。

(2) 一些研究人员认为，生命有可能开始于深海的热液喷口周围。

(3) 以市场为导向的经济复苏政策的一个重大失败就是货币体系改革的失败。

(4) 在该国，这种疫病似乎是地方病，监视和防治活动迄今尚未能成功地阻断病毒在各省之间传播。

(5) 而现实是，谷歌拥有这样一个信息宝藏是一回事，而知道如何利用这个宝藏又是另外一回事。

(6) 泥沙沉积使河道及河口淤积，泄流能力下降，因此洪涝灾害的危险越来越严重。

(7) 溶液中的气体会由浓度高的部分扩散到浓度低的部分。

(8) 每一个签约国承担的主要义务是制定一个国家策略（或行动计划）来管理和保护他们自己的生物多样性。

(9) 印度尼西亚位于太平洋、欧亚大陆和澳大利亚构造板块边缘，正因为这样，它频繁被地震袭击。

(10) 森林砍伐还能改变土壤动态，增加土壤侵蚀，这两者都能把更多的碳释放到大气中。

(11) 一个地区的微量元素浓度差异与气候状况、土地利用程度密切相关。

(12) 该研究团队观察了所谓的底栖生物化石，比如生活在海底的海星、蛤蚌和珊瑚。

(13) 在开始交易前，你应了解适用的规则和伴随的风险。

(14) 这些泥浆和气体由断层处渗出，最后由地表喷出，形成泥火山。

(15) 珊瑚礁被列于世界上最丰富的生态系统之中，可以供养包括珊瑚、鱼和软体动物在内的大批不同种类的海洋生物。

Ⅴ. Exercise 2: Passages in Focus　练习 2：短文翻译

1. Translate the following passage into Chinese

Once a mineral deposit has been discovered in the ocean, the questions concerning whether there is a use for the mineral (a commodity that can be traded) and whether or not technology exists to recover the valuable products from the deposits have to be addressed. This is the starting point in a sequential process that progressively attempts to determine if the resource constitutes a reserve of the mineral in question. The sequential process starts with large seabed areas that are regionally appraised for the deposits that they contain, in order to select the most promising areas for further investigation. This phase is generally described as the prospecting phase. The next phase, exploration, is conducted with a view to identifying deposits, characterizing the deposits, and determining whether or not technology exists or has to be developed to recover the valuable products from the deposits. Where such technology is readily available and tested, the process quickly lends itself to pilot mining tests to establish the economic prospects of mining these deposits. For marine minerals on the continental shelf for which there has been a relatively long tradition of mining, the progression from discovery to production is fairly straightforward. Technology development is generally focused on technology adaptation.

The development of mineral resources of the international seabed area however presents a number of challenges. These include the high costs of exploration (time and money), and the fact that none of the minerals for which rules, regulations and procedures are being developed (polymetallic nodules, polymetallic seafloor massive sulphides and cobalt-rich ferromanganese crusts) have ever been mined in the area. Applied science will have to assist in developing lower cost methods for exploration.

For future mining, in particular in the case of polymetallic nodules, science is currently playing a significant role in the work of the International Seabed Authority with respect to the protection and preservation of the marine environment. For example, in the Clarion Clipperton fracture zone, where the Authority has issued exploration contracts to six entities, projects are currently underway to encourage standardization in the collection of environmental data, including the taxonomic classification of fauna, the determination of species ranges, etc. for the zone. The results of these projects are expected to assist the Authority and contractors to manage impacts from nodule mining. The major challenge however still remains mining technology. In this regard, in the absence of new developments in mining technology to replace the configurations of technology that were proposed in the 1970s and 1980s, discussions of when seabed mining of polymetallic nodules will occur will continue to focus on price increases in the valuable elements that they contain.

2. Translate the following passage into English

从20世纪60年代起，人们对于地球运转方式的了解发生了一次科技革命，大大拓展了人们对海洋矿产的认识。在这次科技革命中，人们对洋盆和大陆的认识发生了重大的改变。在此之前，洋盆被视为巨大的浴缸，被动地容纳着大洋。大陆和洋盆被认为是地球的永恒特征，在地球历史上的大部分时间一直处于现在的状态。

《联合国海洋法公约》有关海洋矿产的条款承认这些非燃料的海洋矿藏，它们来自陆地的侵蚀，以颗粒物或溶解物的形式通过河流被带入海洋。这些矿物包括重金属矿藏，宝石（尤其是钻石），沉积在大陆边缘沉积物中的沙子和砾石，坚硬的岩石基质上的磷灰石，以及沉积在深海底部的多金属结核，这些结核是由溶解在海水中的金属形成的。

这一科技革命建立在板块构造理论的基础之上，它改变了我们对于洋盆的看法——从巨大的浴缸到具有动态的特征，在亿万年的时间里，它随着被称为大陆漂移的陆地运动而或分或合。这一科技革命将洋盆视为多种非燃料矿藏的来源，这些新发现的海洋矿产资源种类中包括大型多金属硫化矿，其中包含数量不等的铜、铁、锌、银、黄金和其他金属。多金属硫化物矿藏是由海底热泉经过成千上万年的时间形成的，这些海底热泉沿着横贯世界各大洋盆的水下活火山山脉而分布。

另外一种新发现的海洋矿产资源是富钴结壳。源自于河流和海底热泉中的金属，在海水中溶解，经过数百万年的时间，沉淀于水下休眠火山的侧面，形成富钴结壳。这些新发现的海洋矿藏都不是可再生资源，因为它们要经过几千年乃至几百万年的时间才能积累到有经济意义的等级和数量。它们是未来的资源，现在并没有投入生产。

在探索海洋方面，我们依然处于早期阶段，详细了解的海底面积只有百分之几，对于海底下面的了解更少，因此今后会继续有新的发现。

Unit Four Marine Pollution

第四单元 海洋污染

Ⅰ. Warm-up for Theme-related Words/Expressions 词汇预习

acidify *v.* 酸化
adhere to 黏附
algae *n.* 藻类
anoxic *a.* 缺氧的
apparatus *n.* 仪器；器械；装置
aquatic *a.* 水生的；水产的 *n.* 水生动植物
ballast *n.* 镇流器；压舱物，压载物
benthos *n.* 海底生物
biodegrade *v.* （进行）生物递降分解；自然降解
bucket system 斗式系统
catch *n.* 捕获量
clog *v.* 阻塞，阻碍；闭合
cobalt *n.* 钴
copepods *n.* 桡脚类动物
coral polyps 珊瑚虫
corrosion *n.* 侵蚀，腐蚀
dairy *a.* 乳制品；奶品 *n.* 乳品业；牛奶场
depletive *a.* 使干涸的；使折损的
diffuse *a.* 四散的，散开的
digestive tract 消化道
discharge *n.*/*v.* 流出；排放
disintegrate *v.* （使某物）碎裂，崩裂，瓦解
dissolution *n.* 溶解，融化
dugong *n.* 儒艮属；儒艮，海牛
ecosystem *n.* 生态系统
entangle *v.* 使纠缠，缠住；使卷入
estuary *n.* 港湾；（江河入海口）河口
eutrophication *n.* 富营养化；超营养作用
exponentially *adv.* 以指数方式
facilitate *v.* 助长；促成
flotsam *n.* 废料；漂流残骸
flux *n.* 流量；流出；熔解；不断的变动
hydraulic pump 液压泵
hydrolysate *n.* 水解产物；酶解物
hydrothermal vents 热液喷口
hypoxic *a.* 含氧量低的
jeopardize *v.* 危及；损害
larvae *n.* 鱼崽
mercury *n.* 水银，汞
mitigation *n.* 减缓；减轻；平静

mutation *n.* 突变，变异
nitrogen *n.* 氮
oscillation *n.* 振动；波动；动摇；〈物〉振荡
periphery *n.* 边缘地带；外围边界
phosphorus *n.* 磷
photodegrade *v.* （使）光致分解；（使）光降解
plankton *n.* 浮游生物
polycyclic aromatic hydrocarbons (PAHs) 多环芳烃
polymers *n.* 高分子化合物；聚合物
polymetallic nodules 多金属结核
precipitation *n.* 沉淀；（雨等）降落；降雨量
residue *n.* 残余；残渣；残留物
retrieval *a.* 回收的；提取的
sediment *n.* 沉淀物；沉渣
stranglehold *n.* 束缚；压制；勒颈
suffocation *n.* 窒息；闷死
sulfide *v.* 硫化物
tailings *n.* 尾矿地；尾矿；尾材；
transoceanic *a.* 跨洋的，在海洋彼岸的
thrive *v.* 茁壮成长；兴盛
turbidity *n.* 混浊度
wreak havoc upon 肆虐；破坏
zebra mussel 斑马贻贝
zinc *n.* 锌
zooplankton *n.* 浮游动物

Ⅱ. Lead-in & Reading 课文导读

Pathways of Marine Pollution

(1) Marine pollution occurs when harmful, or potentially harmful, effects result from the entry into the ocean of chemicals, particles, industrial, agricultural and residential waste, noise, or the spread of invasive organisms.

Eighty percent of marine pollution comes from land. Air pollution is also a contributing factor by carrying off pesticides or dirt into the ocean.

(2) Land and air pollution have proven to be harmful to marine life and its habitats. Nutrient pollution, a form of water pollution, refers to contamination by excessive inputs of nutrients. It is a primary cause of eutrophication of surface waters, in which excessive nutrients, usually nitrogen or phosphorus, stimulate algae growth.

(3) Many potentially toxic chemicals adhere to tiny particles which are then taken up by plankton and benthos animals, most of which are either deposit or filter feeders. In this way, the toxins are concentrated upward within ocean food chains. Many particles combine chemically in a manner highly depletive of oxygen, causing estuaries to become anoxic.

(4) When pesticides are incorporated into the marine ecosystem, they quickly become absorbed into marine food webs. Once in the food webs, these pesticides can cause mutations, as well as diseases, which can be harmful to humans as well as the entire food web.

(5) Toxic metals can also be introduced into marine food webs, which can cause a change to tissue matter, biochemistry, behavior, reproduction, and suppress growth in marine life. Also, many animal feeds have a high fish meal or fish hydrolysate content. In this way, marine toxins can be transferred to land animals, and appear later in meat and dairy products.

(6) Pollution is often classed as point source or nonpoint source pollution[1]. Point source pollution occurs when there is a single, identifiable, and localized source of the pollution. An example is directly discharging sewage and industrial waste into the ocean. Nonpoint source pollution occurs when the pollution comes from ill-defined and diffuse sources. These can be difficult to regulate. Agricultural runoff and wind-blown debris and dust are prime examples.

(7) There are many different ways to categorize, and examine the inputs of pollution into our marine ecosystems. One common path is rivers. The evaporation of water from oceans exceeds precipitation. The balance is restored by rain over the continents entering rivers and then being returned to the sea. The Hudson in New York State, for example, emptying at the northern ends of Staten Island, is a source of mercury contamination of zooplankton

(copepods) in the open ocean. The highest concentration of mercury in the filter-feeding copepods is not at the mouths of these rivers but 70 miles south, nearer Atlantic City, because water flows close to the coast and it takes a few days before toxins are taken up by the plankton. The following are some other pathways of entry by contaminants into the sea.

Direct Discharge

(8) Pollutants enter rivers and the sea directly from urban sewerage and industrial waste discharges, sometimes in the form of hazardous and toxic wastes.

(9) Inland mining for copper, gold, etc., is another source of marine pollution. Most of the pollution is simply soil, which ends up in rivers flowing to the sea. However, some minerals discharged in the course of the mining can cause problems, such as copper, a common industrial pollutant, which can interfere with the life history and development of coral polyps. Mining has a poor environmental track record. According to the United States Environmental Protection Agency, mining has contaminated portions of the headwaters of over 40% of watersheds in the western continental US. Much of this pollution finishes up in the sea.

(10) Marine debris is mainly discarded human rubbish which floats on, or is suspended in the ocean. Eighty percent of marine debris is plastic which presents dangers to wildlife and fisheries in various ways. Many animals living on or in the sea consume flotsam by mistake, as it often looks similar to their natural prey. Plastic debris, when bulky or tangled, is difficult to pass, and may become permanently lodged in the digestive tracts of these animals. Fishing nets made of plastics can be left or lost in the ocean by fishermen. Known as ghost nets, these entangle fish, dolphins, sharks, dugongs, crabs, and other creatures, restricting movement, causing starvation, laceration and infection, and, in those that need to return to the surface to breathe, suffocation.

(11) Plastics accumulate because they don't biodegrade in the way many other substances do. They will photodegrade on exposure to the sun, but they do so properly only under dry conditions, and water inhibits this process. In marine environments, photodegraded plastic disintegrates into ever smaller

pieces while remaining polymers, even down to the molecular level. When floating plastic particles photodegrade down to zooplankton sizes, jellyfish attempts to consume them, and in this way the plastic enters the ocean food chain.

Land Runoff

(12) Surface runoff from farming, as well as urban runoff and runoff from the construction of roads, buildings, ports, channels, and harbors, can carry soil and particles laden with carbon, nitrogen, phosphorus, and minerals. This nutrient-rich water can cause fleshy algae and phytoplankton to thrive in coastal areas; known as algal blooms, which have the potential to create hypoxic conditions by using all available oxygen.

(13) Polluted runoff from roads and highways can be a significant source of water pollution in coastal areas. About 75% of the toxic chemicals that flow into Puget Sound are carried by stormwater that runs off paved roads and driveways, rooftops, yards and other developed land.

Atmospheric Pollution

(14) Another pathway of pollution occurs through the atmosphere. Wind-blown dust and debris are blown seaward from landfills and other areas. For instance, dust from the Sahara moving around the southern periphery of the subtropical ridge moves into the Caribbean and Florida during the warm season as the ridge builds and moves northward through the subtropical Atlantic. There is a large variability in dust transport to the Caribbean and Florida from year to year; however, the flux is greater during positive phases of the North Atlantic Oscillation[2]. The USGS[3] links dust events to a decline in the health of coral reefs across the Caribbean and Florida, primarily since the 1970s.

(15) Climate change is raising ocean temperatures and raising levels of carbon dioxide in the atmosphere. The oceans, as a natural carbon sink, absorb carbon dioxide from the atmosphere. So the rising levels of carbon dioxide are acidifying the oceans. The potential consequences of ocean acidification are not fully understood, but there are concerns that structures made of calcium carbonate may become vulnerable to dissolution, affecting

corals and the ability of shellfish to form shells, that acidification may alter aquatic ecosystems and modify fish distributions, with impacts on the sustainability of fisheries and the livelihoods of the communities that depend on them. Healthy ocean ecosystems are also important for the mitigation of climate change.

Ship Pollution

(16) Ships can pollute waterways and oceans in many ways. Oil spills can have devastating effects. While being toxic to marine life, polycyclic aromatic hydrocarbons (PAHs)[4], found in crude oil, are very difficult to clean up, and last for years in the sediment and marine environment. Oil spills are probably the most emotive of marine pollution events. However, while a tanker wreck may result in extensive newspaper headlines, much of the oil in the world's seas comes from other smaller sources, such as tankers discharging ballast water from oil tanks used on return ships, leaking pipelines or engine oil disposed of down sewers.

(17) Discharge of cargo residues from bulk carriers is another problem. In many instances vessels intentionally discharge illegal wastes despite foreign and domestic regulation prohibiting such actions. It has been estimated that container ships lose over 10,000 containers at sea each year (usually during storms). Ships also create noise pollution that disturbs natural wildlife.

(18) And ballast water taken up at sea and released in port is a major source of unwanted exotic marine life. It is believed that one of the worst cases of a single invasive species causing harm to an ecosystem can be attributed to a seemingly harmless jellyfish. Mnemiopsisleidyi (淡海栉水母), a species of comb jellyfish spreads so quickly that it now inhabits estuaries in many parts of the world. It was first introduced in 1982, and thought to have been transported to the Black Sea in a ship's ballast water. The population of the jellyfish shot up exponentially and, by 1988, it was wreaking havoc upon the local fishing industry. Now that the jellyfish have exhausted the zooplankton, including fish larvae whose numbers have fallen dramatically, yet they continue to maintain a stranglehold on the ecosystem.

(19) Invasive species can take over once occupied areas, facilitate the spread of new diseases, introduce new genetic material, alter underwater

seascapes and jeopardize the ability of native species to obtain food.

Deep Sea Mining

(20) Deep sea mining is a relatively new mineral retrieval process that takes place on the ocean floor. Ocean mining sites are usually around large areas of polymetallic nodules or active and extinct hydrothermal vents at about 1,400 - 3,700 meters below the ocean's surface. The vents create sulfide deposits, which contain precious metals such as silver, gold, copper, manganese, cobalt, and zinc. The deposits are mined using either hydraulic pumps or bucket systems that take ore to the surface to be processed. As with all mining operations, deep sea mining raises questions about environmental damages to the surrounding areas.

(21) Because deep sea mining is a relatively new field, the complete consequences of full scale mining operations are unknown. However, experts are certain that removal of parts of the sea floor will result in disturbances to the benthic layer, increased toxicity of the water column and sediment plumes[5] from tailings. Removing parts of the sea floor disturbs the habitat of benthic organisms, possibly, depending on the type of mining and location, causing permanent disturbances. Aside from direct impact of mining the area, leakage, spills and corrosion would alter the mining area's chemical makeup.

(22) Among the impacts of deep sea mining, sediment plumes could have the greatest impact. Two types of plumes occur: near bottom plumes and surface plumes. Near bottom plumes occur when the tailings are pumped back down to the mining site. The floating particles increase the turbidity, or cloudiness, of the water, clogging filter-feeding apparatuses used by benthic organisms. Surface plumes cause a more serious problem. Depending on the size of the particles and water currents the plumes could spread over vast areas. The plumes could impact zooplankton and light penetration, in turn affecting the food web of the area.

Notes:

[1] point source or nonpoint source pollution　点源污染或非点源污染（面源污染）。美国环保署将点源污染定义为“任何由可识别的污染源产生的污染，‘可识别的污染源’包括但不限于排污管、沟渠、船只或者烟囱”。非点源污染是相对点源污染而言的，是指溶解的和固体的污染物从非特定的地点排出造成的污染。点源污染与非点源污染都有广义和狭义

的两种理解，广义指各种有或没有固定排污口的环境污染，狭义通常限定于水环境的污染。

[2] the North Atlantic Oscillation (NAO)　北大西洋涛动。1920 年英国数学家和气象学家吉尔伯特·沃克爵士发现，北大西洋上两个大气活动中心（冰岛低压和亚速尔高压）的气压变化为明显负相关。当冰岛低压加深时，亚速尔高压加强，当冰岛低压填塞时，亚速尔高压减弱。沃克称这一现象为北大西洋涛动。

[3] USGS (United States Geological Survey)　美国地质勘探局，又名美国地质调查局，是美国内政部所属的科学研究机构。负责对自然灾害、地质、矿产资源、地理与环境、野生动植物信息等方面的监测、收集、分析，对自然资源进行全国范围的长期监测和评估，为决策部门和公众提供广泛、高质量、及时的科学信息。

[4] polycyclic aromatic hydrocarbons (PAHs)　多环芳烃，是煤、石油、木材、烟草、有机高分子化合物等有机物不完全燃烧时产生的挥发性碳氢化合物，是重要的环境和食品污染物。

[5] sediment plumes　沉淀物羽流。羽流（流体力学专业用语），是指一柱流体在另一种流体中移动（a column of one fluid moving through another）。此处是指深海采矿挖掘时造成的尾矿（通常为细小颗粒）被扔回大海后形成了粒子云，漂浮在洋面或悬浮在海底附近的区域。

Ⅲ. Teaching Focus (4): Amplification　翻译技巧（4）：增译

所谓增译，就是在原文的基础上添加必要的单词、词组、分句甚至完整句，从而使得译文在语法、语言形式上符合自身表达习惯，同时在内容、形式和精神等方面上又能与原文对等起来。增译的原则是所增加的词必须在语义、语言结构或修辞上是必不可少的，而不是随心所欲的任意增词。

增译是海洋英语文献翻译常用的方法之一，具体进行英汉互译实践时，大致可以分为以下几种情况：

1. 根据不同语言的表达习惯增译

【例 1】 A new kind of spawning pond—small, cheap, efficient—is attracting increasing attention.

【译文】 一种新型产卵池正在越来越引起人们的注意，这种产卵池体积小，造价低，效率高。

【分析】 按照汉语的语言结构和表达习惯，此句中的形容词“small, cheap, efficient”在英译汉时必须加上一些范畴词，才能更清楚更具体地表达出原文的意思，所以增译了“体积”“造价”和“效率”。另外，因为句子翻译的顺序发生了变化，为了使后半句在语法结构上更加完整，还增译了“这种产卵池”作为后半句的主语，从而使整个译句更加符合汉语的表达习惯。

【例 2】The pathway of fatty acid biosynthesis in fish is known to be basically similar to that operating in mammals.

【译文】人们知道，鱼类的脂肪酸合成途径与哺乳动物基本相似。

【分析】英语中被动句出现的频率远远高于汉语，在英译汉时，常常会将其译为主动句。这样就需要增加行为主体。本句翻译增译了“人们知道”，其目的是使句子语言结构更加完整，符合汉语语法表达习惯。

同样是因为英汉语言表达习惯的差别，当我们进行汉译英时，需要进行反向思维，即将汉语主动句转译为英语被动句。这一翻译技巧在科技文献翻译时尤为常见。被动句的频繁使用是英语科技文体的一大特征，它可以避免内容的主观感，准确传达出相关研究方法、数据或结论等的科学性。在进行汉译英时，因为主动转成了被动，所以常常需要根据具体情况来增译一个动词或一个主句，以保持句子在语法结构上的完整。例如：

【例 3】森林发展和养护是防止土地退化、保护物种和增加碳螯合的重要手段。

【译文】Forest development and conservation were seen as important means to combat land degradation，preserve species and sequester carbon.

【分析】本例将汉语主动句转为英语被动句时，增译了“seen as”这个动词词组，更契合英语的语感。

【例 4】现在黄海每年向东海输运悬浮沉积物仅为 $0.2\times10^8\sim0.3\times10^8$ 吨，主要是废黄河沿岸及水下三角洲受侵蚀再悬浮的黄河泥沙。

【译文】It has been shown that only $0.2\times10^8-0.3\times10^8$ t suspended particles are carried to the East China Sea annually. They are chiefly resuspended sediment from erosion of the coast and subaqueous delta of the abandoned Yellow river.

【分析】本例增译一个由“It”作为形式主语引导的被动句句式结构。“It has been shown that”充当整个复合句的主句部分，增强了语言的客观性，使句子更符合科技英语的表达习惯。另外，此句翻译时还进行了分译处理，由汉语的一句话译成了英语的两句话。为了保证第二个句子结构完整，同时使前后两句读起来关联性更强，增译了主语代词“They”。

2. 根据上下文语义和逻辑关系增译

【例 5】Hofer（1982）also calculated that the daily production of proteolytic enzymes in roach feeding on grass was lighter than in those fed mealworms.

【译文】霍弗（1982）还进行了计算，结果表明，拟鲤在摄食草类时，蛋白水解酶的日产量比摄食粉蛀虫时高。

【分析】译句中增加了“结果表明”，可以使句义及上下逻辑关系更加清晰，前后衔接和过渡得更加自然。

【例 6】Nonsense termination in human cells may also be susceptible to suppression by antisense agents，providing a new approach to address numerous diseases caused by nonsense mutations.

【译文】由于人体细胞的无义突变也可能受到反义试剂抑制作用的影响，这就为治疗由无义突变引起的多种疾病提供了一个新途径。

【分析】根据上下文的语义关系，原句的前半句所陈述的事实是形成后半句结论的依据。所以在翻译时，前半句增加了表示依据关系的连接词“由于”，后半句增加表示承接意义的“这就”，使得整个译句前后连贯，逻辑关系一目了然。

【例 7】Under the sunlight irradiation，semiconductor phtocatalytic materials have the ability to split water，photodegrade organic pollution，photo-reduce carbon dioxide and so on. They are considered to have great potentials in solving the problem of energy shortage and environmental pollution.

【译文】半导体光催化材料在阳光的照射下具有光分解水质氢、光降解有机污染物、光还原二氧化碳等功能，因此其在解决能源短缺和环境污染方面具有巨大的应用前景。

【分析】本例原文有两句，两句之间有着明显的因果关系。翻译时，增译表示因果关系的连词“因此”，成功地将原文的两句并译为一句，既明确了上下文语义上的因果关系，又达到了使译文更加精炼简洁的目的。

【例 8】In the method material cost is fall，full utilizing industry flotsam，making flotsam into usefulness substance accord with environment requirements and it inaugurate a new route that utilize industry solid flotsam with reason.

【译文】这样既降低原料成本，又能充分利用工业废物、变废为宝，既符合环保要求，又为合理利用工业固体废料开辟了一条绿色环保的途径。

【分析】本例中的译文运用两个“既…又…”句式，厘清了原文中的几个并列结构之间的关系，同时使译文排比整齐，达到了很好的修辞效果。

3. 根据原文文中蕴含的意思进行增译

【例 9】Turbidity value and bacteriological pollution index could reach national drinking water first level standard by potassium ferrate at the test condition.

【译文】在本试验条件下经高铁酸钾处理后的水源水，其浊度和细菌学指标能满足国家饮用水一级标准。

【分析】译句中画线部分文字在原文中没有对应的具体表述，但是其意思已经蕴含在句意之中。根据学科背景知识，可知高铁酸钾通常用于处理或净化水，且句中还提到了“drinking water”，所以将介词词组“by potassium ferrate”明确地译成了“经高铁酸钾处理后的水源水”。通过增译，使得译文语义更加清楚，有效避免了歧义。

【例 10】The most restrictive MPAs are the marine reserves, also known as“ecological” reserves or “no-take” reserves, marine reserves are special class of MPA.

【译文】最具限制性的海洋保护区是海洋自然保护区，也称为“生态”自然保护区或“禁捕”自然保护区，自然保护区是一类特殊的海洋保护区。

【分析】“reserves”本义为“保护区”，在翻译时，增加了“自然”一词进行限定，译为“自然保护区”。因为保护区可以有多种类型，自然保护区是海洋保护区中非常特殊的一类，虽然原文中没有特意说明，其所指却已经蕴含在上下文之中，所以有必要精确地翻译出来，以区别于其他类型的保护区。

【例 11】The ocean nepheloid layers are both the pathway for transporting the terrigenous materials to the seabed and the residence place of sinking and resuspended particulate matters.

【译文】海洋雾状层既是陆源物质进入海底的输送通道，又是海洋水体中沉降颗粒及底部再悬浮颗粒物的停留场所。

【分析】同样根据学科背景知识可知，海洋雾状层是发生在海洋水体中的一种现象，虽然原句中没有明确指出，但是很有必要翻译出来，所以后半句增译了“海洋水体中”。另外，译句中还重译了“particulate matters”。

【例 12】The results demonstrated that the surface wetted radius and vertical wetted depth were exponentially proportional to the applied water volume.

【译文】研究结果表明，径向和垂直湿润距离随灌水量的增加呈幂函数关系增加。

【分析】本例原文中没有“增加”的对应词，但是根据常识，此处的按比例变化的关系应该是“增加”的关系。

4. 根据不同语言的语法规则进行增译

【例 13】One suggested requiring mercury producers to reprocess the residues and wastes from the mercury that they supplied；another stressed the need to consider illegal trade in mercury.

【译文】一名发言者建议要求汞的生产商对来自其所供应汞的残渣和废物进行再加工；另外一名发言者则强调有必要审议非法汞贸易的问题。

【分析】英语的“one”和“another”都有代名词的功能，其具体所指示的名词往往蕴含在上下文中。而汉语的数词没有代名词功能，所以英译汉时，需要译出具体的名词。事实上，此句除了增译了名词“发言者”之外，还增译了量词“名”。这是因为汉语的量词非常丰富，汉语的表达习惯往往是数词量词同行的；而且需要根据不同的名词搭配不同的量词使用，而此处名词“发言者”需要跟量词“名”来搭配使用。

【例 14】人类为了经济利益，对自然进行了无限度的开发，给生态环境带来了严重的破坏并威胁和影响着社会的稳定和发展。

【译文】The humanity，in order to pursue the economic interest，has carried on non-limit claim to the nature and the depletive development，has brought the serious destruction for the ecological environment and threatens and is affecting the society's stability and development.

【分析】因为英语的“in order to”是一个不定式结构，其后必须跟一个动词，所以翻译时增加了动词“pursue”，与原文中的“经济利益”形成语法和语义上的搭配。同时，为了更加精确地表达出原文的意思，在翻译“无限度的开发”时，进行了重译处理，通过“non-limit claim to the nature”和“depletive development”两个部分表达出来，这样增译处理的句子，既符合了词与词之间的搭配习惯，又照顾到了意思的完整。

【例 15】The agenda they are putting forward is laden with what sounds like low-hanging fruit and the trick is getting any of those apples to fall from their trees.

【译文】他们提的这个计划的未来听起来似乎会硕果累累，不过关键还是在于如何让果实从树上掉下来。

【分析】本例增译“如何”更加符合汉语的语法规则。

【例 16】Mussels can thrive in polluted water because of an inborn ability to purify bacteria，fungi，and viruses.

【译文】蚌类可以在受污染的水中生活，因为它们天生有净化细菌、真菌和病毒的能力。

【分析】英文中的“because of”是介词词组，后面必须接用代词、名词或动名词做宾语。而汉语中的“因为”是连词，为了让后半句在语法结构上更加完整，同时使上下文语义连贯和明确，此处增译了主语“它们”。

5. 根据词汇的语义搭配进行增译

【例 17】Appetite and satiation are of importance to fish farmers because of the need to ensure that feed regimes（feeding frequencies，ration sizes and the time over which they are dispensed）are adjusted to maximize consumption，growth and conversion efficiencies.

【译文】由于鱼类养殖者需要调整投喂方式（投喂次数、投喂量和每次投喂所持续的时间）以确保摄食量最大，生长速度最快，饲料转化效率最高，所以，食欲和饱食对鱼类养殖者十分重要。

【分析】“maximize”为及物动词，意思是“使…最大化”；在原文中，其后共接了“consumption”（摄食）、“growth”（生长）和“conversion efficiencies”（转化效率）三组名词做宾语，在英语中这样搭配使用完全正常。根据汉语修辞和搭配的需要，译句中将三个名词分别增译为“摄食量”“生长速度”和“饲料转化效率”；同时，将“maximize”分别译为“最大”“最快”和“最高”，以符合汉语“量大”“速度快”和“效率高”的词汇和语义搭配习惯。

【例 18】The excess of oxygen depleting chemicals can also lead to hypoxia and severe reductions in water quality，fish，and other animal populations.

【译文】耗氧的化学物质过量也会导致水体缺氧，水质急剧下降，以及鱼类和其他动物种群数量的大幅度减少。

【分析】与【例 17】同理，本例中原句的“severe reductions”分别译为“急剧下降”和“大幅度减少”，这也是由汉语的修辞和搭配习惯决定的。

综上所述，增译是英汉互译实践中非常常见的一种翻译技巧。但有一点必须格外慎重对待，即增译必须同时兼顾“忠实”的翻译标准。增译绝对不是任意添加新意。增译的目的是使译文更加通顺、意义更加明确，使其表达方式在语法规则、语义搭配、语感等方面更加符合目标语的习惯。

Questions for Discussion:

1. How to strike the right balance between amplification and faithfulness in translation?

2. Do you think the application of amplification in Chinese-English translation is just the same as in English-Chinese translation? Why or why not?

Ⅳ. Exercise 1：Sentences in Focus　练习 1：单句翻译

1. Translate the following sentences into Chinese

(1) Nutrient pollution，a form of water pollution，refers to contamination by excessive input of nutrients. It is a primary cause of eutrophication of surface water，in which excessive nutrients，usually nitrogen or phosphorus，stimulate algae growth.

(2) Many potentially toxic chemicals adhere to tiny particles which are then taken up by plankton and benthos animals，most of which are either deposit or filter feeders.

(3) Toxic metals can also be introduced into marine food webs，which can cause a change to tissue matter，biochemistry，behavior，reproduction，and suppress growth in marine life.

(4) The highest concentration of mercury in the filter-feeding copepods is not at the mouths of these rivers but 70 miles south，nearer Atlantic City，because water flows close to the coast and it takes a few days before toxins are taken up by the plankton.

(5) However，some minerals discharged in the course of the mining can cause problems，such as copper，a common industrial pollutant，which can interfere with the life history and development of coral polyps.

(6) Plastic debris，when bulky or tangled，is difficult to pass，and may become permanently lodged in the digestive tracts of these animals.

(7) In marine environments，photodegraded plastic disintegrates into ever smaller pieces while remaining polymers，even down to the molecular level.

(8) Known as ghost nets，these entangle fish，dolphins，sharks，dugongs，crabs，and other creatures，restricting movement，causing starvation，laceration and infection，and，in those that need to return to the surface to breathe，suffocation.

(9) This nutrient-rich water can cause fleshy algae and phytoplankton to thrive in coastal areas；known as algal blooms，which have the potential to create hypoxic conditions by using almost all available oxygen.

(10) Dust from the Sahara moving around the southern periphery of the

subtropical ridge moves into the Caribbean and Florida during the warm season as the ridge builds and moves northward through the subtropical Atlantic.

(11) The potential consequences of ocean acidification are not fully understood, but there are concerns that structures made of calcium carbonate may become vulnerable to dissolution, affecting corals and the ability of shellfish to form shells.

(12) However, while a tanker wreck may result in extensive newspaper headlines, much of the oil in the world's seas comes from other smaller sources, such as tankers discharging ballast water from oil tanks used on return ships, leaking pipelines or engine oil disposed of down sewers.

(13) It is believed that one of the worst cases of a single invasive species causing harm to an ecosystem can be attributed to a seemingly harmless jellyfish.

(14) However, experts are certain that removal of parts of the sea floor will result in disturbances to the benthic layer, increased toxicity of the water column and sediment plumes from tailings.

(15) The floating particles increase the turbidity, or cloudiness, of the water, clogging filter-feeding apparatuses used by benthic organisms.

2. Translate the following sentences into English

(1) 塑料袋不会自然降解，它们会光裂解成越来越小的有毒碎片，持续污染着土地和水道；如果被动物误食，则进入食物网中。

(2) 结果表明，这种活性成分物质能溶于水，不耐酸，不耐加热，可被乙醇钝化，属蛋白质类物质。

(3) 不仅如此，物质本身也会分解：长期以来被认为完全稳定的质子也许在大约 1 039 年后会裂解。

(4) 它们浮在小船和残骸上，被水流拖入内陆水域，直到水流开始逆向涌动，将它们带回海床。

(5) 随着越来越多的工程（如水电、采矿、核废料掩埋及溶质运移等）修建在岩体之上或岩体之中，地下水问题摆在了我们面前，它对工程安全起着重要作用。

(6) 它们在宿主细胞中大量复制，持续地涌入周围环境中，以至于如果将全球海洋中漂浮着的所有病毒性物质收集起来，其总重能超过所有蓝鲸的重量。

(7) 将农作物废料和其他生物物质转化成炭并将其施于热带土壤的方法，

可以固化碳并能提高农作物产量。

(8) 论文采用离子浓度分析及浊度分析的方法进行了天然海水与地层水的静态配伍性研究。

(9) 虽然温室效应的总体效应仍是不可知的，但全球变暖导致了两极冰盖融化，海平面上升，直接危害到了我们的生态系统。

(10) 到了去年三月，一种流感病毒的变种突然出现了，科学家们知道它有肆虐全国的可能，但为时已晚。

(11) 在该地区，淡水养殖和海水养殖的“蓝色革命”正在以指数方式飞速发展。

(12) 微藻类靠吸收二氧化碳生长，可以作为家畜饲料和生物燃料以及制造塑料的原材料。

(13) 它在对云层、降水、冰雹和雷暴的气象探测以及飞机和船舰的导航中，起着日益重要的作用。

(14) 在生态环境日益恶化，资源日益枯竭的今天，人们越来越重视环境问题，以及经济、环境和社会的可持续发展。

(15) 我们连续四年在全国范围内开展整治违法排污企业的活动，以确保群众健康和环境安全。

Ⅴ. Exercise 2：Passages in Focus　练习 2：短文翻译

1. Translate the following passages into Chinese

Eutrophication is a common phenomenon in coastal waters. The World Resources Institute has identified 375 hypoxic coastal zones around the world, concentrated in coastal areas in Western Europe, the Eastern and Southern coasts of the US, and East Asia, particularly in Japan. In contrast to freshwater systems, nitrogen is more commonly the key limiting nutrient of marine waters; thus, nitrogen levels have greater importance to understanding eutrophication problems in salt water.

One proposed solution to eutrophication in estuaries is to restore shellfish populations, such as oysters. Oyster reefs remove nitrogen from the water column and filter out suspended solids, subsequently reducing the likelihood or extent of harmful algal blooms or anoxic conditions. Filter feeding activity is considered beneficial to water quality by controlling phytoplankton density and sequestering nutrients, which can be removed from the system through shellfish harvest, buried in the sediments, or lost through denitrification. Foundational work towards the idea of improving marine water quality through shellfish cultivation was conducted by using mussels in Sweden.

One typical example is the "bloom" or "red tide" which is great increase of phytoplankton in a water body as a response to increased levels of nutrients. Scientists prefer the phrase "harmful algal bloom" (HAB) to "red tide" because blooms are not always red and are not related to the tides. In the ocean, there are frequent HABs that kill fish and marine mammals and cause respiratory problems in humans and some domestic animals when the blooms reach close to shore. Every U. S. coastal state experiences HABs, which has become a national concern because they affect not only the health of people and marine ecosystems, but also the "health" of their economy—especially coastal communities mainly dependent on the income of jobs generated through fishing and tourism. Coastal HAB events have been estimated to result in economic impacts in the United States of at least $82 million each year.

With climate change and increasing nutrient pollution potentially causing HABs to occur more often and in locations not previously affected, it's important to learn as much as possible about how and why they form and where they are, so that their harmful effects can be reduced. Studies indicate that many algal species flourish when wind and water currents are favorable. In other cases, HABs may be directly linked to "overfeeding". This occurs when nutrients from land sources flow into bays, rivers, and the sea, and build up at a rate that "overfeeds" the algae that exist normally in the environment. Some HABs appear in the aftermath of natural phenomena like sluggish water circulation, unusually high water temperatures, and extreme weather events like hurricanes, floods, and drought. Although many coastal countries experience HABs, different organisms live in different places and cause different problems. Other factors, such as the structure of the coast, runoff, oceanography, and other organisms in the water, can also change the scope and severity of HAB impacts.

2. Translate the following passages into English

海洋和沿海生态系统在全球碳循环中起着重要作用，它们清理了 2000 年至 2007 年间人类活动排放的二氧化碳的 25%和工业革命以来人为二氧化碳排放量的约一半。海洋温度上升和酸化意味着海洋碳汇能力会逐渐减弱，这已引起全球关注。

2008 年 5 月，美国国家海洋和大气管理局的科学家们在《科学》杂志上发表了一份报告。他们发现大量相对酸化的海水正涌向北美太平洋大陆架 4 英里之内的区域。这里是多数本土海洋生物生活或出生的重要地带。虽然报告只提到了从温哥华到加利福尼亚北部区域，但大陆架其他区域可能也有类似情况发生。

与此有关的一个问题是在海床沉积物下发现的甲烷水合物储层。这里圈留着大量的温室气体甲烷。一旦海洋变暖，这些气体就有可能释放。2004 年全球海洋甲烷水合物的存储量预计介于 100 万～500 万立方千米之间。如果所有这些水合物均匀平铺于海底，其厚度可达三至十四米。这个预估量相当于 500 亿～2 500 亿吨碳，与从所有其他化石燃料储量中估计出的 5 000 亿吨碳可以相提并论。

除了酸化之外，海洋中还存有许多其他类型的污染，如噪音污染。海洋生物对过往船只、石油勘探、地震勘测和海军低频主动式声呐等所造成的噪音污染非常敏感。声音在海洋中传播比在大气中更迅速，传播距离也更远。海洋动物，如鲸类，常常视力微弱。它们基本上生活在一个由声学信息定义的世界里。许多生活在黑暗世界里的深海鱼类亦是如此。1950—1975 年间，太平洋任一位置上的环境噪音提高了约十分贝（这意味着其强度提升了十倍）。

噪音亦迫使动物大声交流，这就是隆巴德效应。当潜艇探测器运转时，鲸发出的声波更长。如果海洋动物“说话”不够响亮，他们的声音就可能被人为的声音掩盖。而这些人类听不到的声音可能是海洋动物发出警告或是寻找猎物的声音。当某一物种开始大声交流，就会掩盖其他物种的声音，最终导致整个生态系统物种交流的声音更大。

Unit Five　Marine Disaster

第五单元　海洋灾难

Ⅰ. Warm-up for Theme-related Words/Expressions　词汇预习

accommodate　*v.* 容纳
afflict　*v.* 使受痛苦；折磨
auxiliary　*a.* 辅助的；备用的
Canadian Coast Guard（CCG）　加拿大海岸警卫队
capsize　*v.* 使（船或车）翻；倾覆
cargo　*v.*（船或飞机装载的）货物
casualty　*n.* 伤亡（人数）；事故
catastrophic　*a.* 灾难的；惨重的
Coast Guard　海岸警卫队
Coast Guard Auxiliary（CGA）　辅助海岸警卫队
collision　*n.* 碰撞；冲突
conceivable　*a.* 可想到的，可相信的
contingency　*n.* 应急；偶发事件
coordinator　*n.* 协调员
delegate　*v.* 委派代表；授权给
Department of National Defence（加拿大）国防部
dependant　*a.* 随…而变的
designate　*v.* 指出；指派
errand　*n.* 差事；使命
evacuation　*n.* 撤离；疏散
flexible　*a.* 灵活的；易弯曲的
grounding　*n.* 搁浅
hazard　*n.* 危险；冒险的事
hinder　*v.* 阻碍，妨碍
Incident Command Centre（ICC）　事故指挥中心
Incident Command System（ICS）　事故指挥系统
infectious　*a.* 有传染性的；易传染的
infrequent　*a.* 稀少的；罕见的
interagency　*n.* 跨部门的
interface　*n.* 交接；接口　*v.* 接合；交流
in conjunction with　与…协力
Joint Rescue Co-ordination Center（JRCC）　联合救援协调中心
liaison　*n.* 联络，联络人
Major Maritime Disaster Contingency Plan（MMDCP）　重大海难应急预案
marshalling　*n.* 召集，集结
mass rescue operation（MRO）　大规模救援行动

migrant vessel 移民船只
multi-agency *n.* 多机构
multi-casualty incident（MCI） 重大伤亡事故
non-aligned *a.* 不结盟的；中立的
On-Scene Commander（OSC） 海难现场指挥员
overtax *v.* 使负担过重，使过度疲劳
piracy *n.* 海上抢劫
scene management 现场管理
seamanship *n.* 航海技术；船舶驾驶术
Search and Rescue（SAR） 搜救
shoreline *n.* 海岸线
smuggle *v.* 私运；走私
spill *n.* 泄露；溢出
strand *v.* 搁浅；陷入困境
stretcher *n.* 担架
resilience *n.* 快速恢复的能力
tally *n.* 记录，测量
triage *n.* 医疗类选法
vessel *n.* 容器；船

Ⅱ. Lead-in & Reading 课文导读

Marine Disaster[1] Scene Management[2]

Introduction

(1) What is a disaster? An event, either natural or man-made, that causes great distress or destruction or that requires a response beyond the normal capacities of the agencies involved.

(2) In the marine context this will often be a multi-casualty incident

(MCI) or mass rescue operation (MRO) that requires a multi-agency response.

(3) An important consideration is that the major marine disaster is a relatively low probability but high consequence event. This speaks to the need for training and preparation in anticipation of such an event, as infrequent as it may be. Normal operational methods are less likely to be successful and the consequences of mistakes can be catastrophic.

(4) The principles of marine disaster scene management can also be applied to incidents that may not fit most people's definition of a disaster. It does not take a very large number of distressed or injured individuals to overwhelm a single Coast Guard[3] resource.

(5) Disasters at sea cannot usually be completely resolved by applying disaster plans evolved on land and as such this section presents information that will help you manage a marine disaster. It deals specifically with the responsible agencies and interagency operations, the key Coast Guard roles and the keys to success during a disaster response.

(6) Remember that no plan can ever fully apply to all situations. No disaster plan should ever be cut in stone. One of the biggest challenges in dealing with such an incident is the need to be flexible and to adapt easily to an unfamiliar situation. Knowledge of the principles of marine disaster scene management along with common sense, communication skills and good seamanship will be very important if and when you encounter a multi-casualty situation.

Disaster Requiring a Coast Guard Response

(7) Though relatively infrequent, these high impact and potentially high consequence events do take place. Incidents that have or may occur include, but are certainly not limited to the following:

- Accidents involving large passenger vessels incidents including groundings, collisions, capsize or sinking, fire at sea or in port, chemical spill, etc.
- Emergencies involving offshore exploration or exploitation platforms
- Terrorism or piracy
- Aircraft emergency landing or crash at sea or in tidal area

• Infectious disease outbreaks
• Extreme weather events affecting multiple vessels
• Vessels stranded or crushed in ice
• Migrant vessels or smuggling of human cargo
• Land disaster requiring maritime evacuation

(8) Consider also that Coast Guard resources may well be the first assistance available to a small, isolated community afflicted by some form of disaster. It is conceivable that this assistance may need to be provided with no help from other agencies for some time.

(9) The marine environment may in general make response more difficult since an emergency at sea often involves continued hazards and complications that must be dealt with in addition to the need to care for and evacuate survivors.

Jurisdiction and Responsibility

(10) Jurisdiction and responsibility is dependant on the location and nature of the incident. With respect to the marine disaster the Victoria Search and Rescue Region, Major Maritime Disaster Contingency Plan (MMDCP) describes responsibilities for the Owner/Agent, for the Master of the Distressed Vessel, for the SAR Authorities and the Provincial Authorities. The following applies to SAR Authorities... In accordance with Canadian and International laws and conventions... SAR authorities (DND, CCG) are responsible for:

• Incorporating an Incident Command Centre (ICC) with incorporates the JRCC
• Developing a SAR plan of action...
• Accounting for all persons on board and searching for missing persons
• Rescuing, stabilizing and transporting casualties for treatment
• Rescuing and evacuating survivors to reception centres
• Notifying collateral authorities
• Addressing all media enquiries regarding the SAR operation

(11) In a marine disaster, after safety of responders, SAR of survivors is the first priority. Once the SAR is resolved, then environmental and investigation take over—these not be in conflict. Also, as the incident moves

from the marine environment to shore, the responsibility will be handed from the JRCC to the civil authority.

(12) The Master of the distressed vessel is ultimately responsible for the welfare of his passengers and crew. This includes timely evacuation if required, developing an action plan in conjunction with SAR authorities, and sharing information on all significant developments with SAR authorities. In short the CG is not in charge of evacuation but should assist the Master as appropriate; while the Master is not in charge of the SAR response but consults as appropriate.

Responsible Agencies

(13) As noted, responsibility for control of the incident will depend on location and jurisdiction. Effective incident management depends in part on having a clearly established command that is well communicated to all participating resources/agencies. Communications between agencies is frequently a challenge during a multi-agency response; pre-planning and additional effort in this area during incidents is critical.

Joint Rescue Coordination Centre

(14) The Joint Rescue Coordination Centre, or JRCC, is a DND operation staffed by Canadian Armed Forces and Coast Guard personnel. The JRCC receives reports of an emergency situation and tasks Coast Guard and Canadian Forces resources as required. During a marine or air incident, JRCC is the responsible authority and provides executive control of the operation. JRCC may provide an interface or liaison function with the provincial emergency program, ambulance authorities, police, any non-aligned resource or service.

Canadian Coast Guard

(15) The Coast Guard and the Coast Guard Auxiliary (CGA) provide the two largest marine elements of the search and rescue organization, and are tasked by the JRCC. The role of the Coast Guard in a multi-casualty situation will depend on where it occurs. If at sea, then triage[4], patient care and evacuation with tracking are likely to be carried out by Coast Guard personnel. If the disaster occurs on a shoreline with other responsible agencies responding, then Coast Guard resources may be tasked as requested to assist

those civil authorities.

US Coast Guard

(16) There is a history of cooperation between US and Canadian Coast Guards, both during exercises and on incidents. This includes the sharing of marine and air resources, as well as mutual support related to incident coordination and communications, particularly in border areas. The USCG uses the Incident Command System (ICS) as its model for command and control during a major incident.

Key Coast Guard Personnel

JRCC Coordinator

(17) The JRCC coordinator will be either a Coast Guard officer in a marine situation or a DND officer for an air incident. Air and marine will support each other in a major incident.

(18) The coordinator initially receives indication of a problem and assigns appropriate resources. During the incident, the coordinator gives direction to the OSC and provides a communications interface between agencies.

On-Scene Commander (OSC)

(19) The OSC will be designated by and represents JRCC at the scene. JRCC will usually select the largest available SAR vessel with the biggest executive structure and superior communications capabilities.

(20) Depending on circumstances, the OSC may delegate the actual operation of the vessel to subordinates in order to be able to concentrate on managing the incident. The OSC must be prepared to step back from active involvement so as not to be overtaxed. It is also appropriate to periodically re-evaluate the status of OSC and reassign this role if in the best interests of managing or resolving the incident.

(21) The OSC should appoint Transport and Triage Officers in anticipation that these functions may be required. These roles may be combined, dependant on resources and situation. As soon as trained rescue personnel are on scene, the OSC or Transport Officer will determine if the scene is STABLE or UNSTABLE, and convey this information to the triage officer and rescue personnel who may approach the scene or board the distressed vessel.

(22) As the incident progresses the OSC serves as the eyes and ears of the JRCC, provides on scene decision making (including stability, coordination of resources, and rescue plan), requests additional resources as required and serves as a vital communications link between JRCC and resources on scene.

Transport Officer

(23) If the OSC is the eyes and ears of the JRCC then the Transport Officer may be thought of as the eyes and ears of the OSC. The Transport Officer role may best be filled by a ship's officer and is required in a major incident or mass rescue operation or if the situation is such that a number of trips need to be made to accommodate all survivors. If resources are limited this role may need to be filled by the Triage Officer or another RS (Rescue specialist) at scene.

(24) The Transport Officer carries out the following duties:

• As appropriate, serves as leader of the rescue party at the incident scene (e. g. on board the stricken vessel, on shore, etc.)

• On behalf of the OSC makes contact with the vessel master, provides an ongoing assessment of scene stability and numbers of POB (Persons on Board)

• Aids in selecting casualty collection areas and most suitable evacuation point(s)

• With Triage Officer assign survivors to the transport in a logical sequence, as determined by triage and scene stability

• Coordinates marshalling of the transport units and requests additional resources if required

• Is responsible for casualty tracking—keeps accurate tally of evacuees, with as much detail as possible

• Periodically updates the OSC as to on scene and transport status

Triage Officer

(25) The Triage Officer should be appointed prior to reaching the disaster scene. He or she must be familiar with the principles of marine disaster scene management and triage. A senior Rescue Specialist on a CG vessel is a good candidate for this role.

(26) Once on scene, the Triage Officer carries out the following duties:

• Reports to the OSC via the Transport Officer (if assigned)

• Aids in determining scene stability, casualty collection area and evacuation points

• Quickly estimates the number and severity of casualties and formulates a triage plan appropriate to the situation (recruit assistance as required)

• Conducts a rapid triage (as with START Triage) and tags each patient to indicate the priority for care or evacuation, as determined by scene stability

• After initial triage, re-evaluates and conducts secondary triage as appropriate

Rescue Party

(27) The members of a Coast Guard rescue party will, after determination of scene stability and hazards, assist the Transport and Triage Officers as required. The specific duties of the rescue party may include:

• Assist with damage control if deemed appropriate

• Accompany Triage Officer and perform critical interventions as directed

• Gather and set up rescue or first aid equipment

• Locate patients and perform basic first aid (urgent first or as directed by Triage Officer) and carry them, by stretcher if necessary to the casualty collection area

• Locate able bodied survivors or bystanders and enlist their help as stretcher bearers or in otherwise assisting with the rescue effort CG Rescue Party carries stretcher during training exercise

• Assist Transport Officer with casualty tracking

Bystanders

(28) Whether a multi-casualty incident occurs on land or on the water, there are usually bystanders. Depending on the actions of the trained rescuers, these people will either hinder or help the operation. Bystanders may be assigned jobs that are within their capability by the rescuers on scene. Such jobs could include:

• Providing pleasure craft as transport from the rescue scene to shore

• Assisting with stretcher handling and basic first aid

• Running errands and providing food and clothing

• Caring for children and offering psychological support

Survivors

(29) Uninjured survivors of a disaster often show amazing resilience and

can sometimes be extremely useful in the rescue effort. Be sure that any survivors who offer to help are uninjured and that they are accounted for. These people can assist by:

- Assisting with stretcher handling and basic first aid
- Comforting other survivors

(30) Try to ensure that uninjured survivors are eventually evacuated along with the others for purposes of proper debriefing, processing and accounting.

Notes:

[1] marine disaster 海难，是指船舶在海上遭遇自然灾害或其他意外事故所造成的危难。海难可给生命、财产造成巨大损失。造成海难的事故种类很多，大致有船舶搁浅、触礁、碰撞、火灾、爆炸、船舶失踪，以及船舶主机和设备损坏而无法自修以致船舶失控等。

[2] scene management 现场管理，是指用科学的标准和方法对现场各要素，包括人（工人和管理人员）、机（设备、工具、工位器具）、料（原材料）、法（加工、检测方法）、环（环境）、信（信息）等进行合理有效的计划、组织、协调、控制和检测，使其处于良好的结合状态，以达到优质、高效、低耗、均衡、安全的目的。在本文中，海难现场管理（marine disaster scene management）指用科学的标准和方法应对海难现场各要素，以达到高效进行海难救助的目的。

[3] Coast Guard 海岸警卫队，也就是水上警察，但它的体制和职能则是警察和武警这两支武装部队所无法比拟的。它是负责沿海水域、航道的执法、水上安全、遇难船只及飞机的救助、污染控制等任务的武装部队。

[4] triage 医疗类选法，是指在伤病者同时多发出现的情况下，以五色的等级颜色区分重伤度及紧急度，根据紧迫性和救活的可能性等决定哪些人优先治疗。

Ⅲ. Teaching Focus (5): Omission 翻译技巧（5）：省译

省译是与增译相对应的翻译方法，是指在不损害原文思想内容的前提下，删去原文中一些多余的或不符合译入语表达方式或习惯的词语，使译文简洁流畅。在同一译例里，如果英译汉用了增译，那么，回译成英语就是省译。一般来说，汉语较英语简练。因此，在英译汉时，如果将许多在原文中必不可少的词语原原本本地译成汉语，就会成为不必要的冗词，译文会显得十分累赘。因此省译在英译汉中使用得非常广泛，其主要目的是删去一些可有可无、不符合译文习惯表达法的词语，例如实词中的代词、动词的省略；虚词中的冠词、介词和连词的省略等。同时，汉语也有重复啰唆、拖泥带水的语言现象，因此省译也同样适用于汉译英的某些情况。汉译英的省略主要是省略冗词赘语，以及一些表示概念范畴的词语和过分详细的描述。海洋英语文献翻译中的省译不能凭空乱减，需要做到“浓缩至精华”的简化。省译在海洋英语翻译中的应用可以从语法、意义和修辞三个方面来看。

1. 语法上需要进行的省译

（1）冠词的省译。英语中有冠词，但汉语中却没有冠词，因此在英译汉时，往往会省略冠词，从而实现译文的简练客观。

【例 1】An understanding of the entire design sequence is essential to anyone seeking to develop a basic design of a submarine.

【译文】想要进行潜水艇的基本设计，就必须了解整个设计顺序。

【例 2】A man without a diving suit will find it difficult or impossible to breathe if he goes down too far.

【译文】如果不穿潜水服而下潜太深的话，人就会发现呼吸困难，或者根本无法呼吸。

【分析】这两个例子都省译了名词前面的不定冠词。通常在准确传达原文内容的基础上，会省译冠词，这体现了海洋英语简洁明确的特点。

（2）连词的省译。英语重形合，而汉语重意合。英语中逻辑关系多由连词表示，而汉语很少借助连词，只是通过语序及语言内涵来表意。一般来说，英语中的连词可以分为并列连词和从属连词两种，后者包括时间连词、地点连词、原因连词、条件连词、程度与结果连词。它们在句子中只是起到连接作用而不具句子成分的功能，以表达句子的逻辑关系。汉语则将时间、逻辑关系暗含在句子中，不需要使用连词。

【例 3】Some ship operating companies are closely tied to successful previous designs，and they will permit little variation from these baselines in the development of replacement vessel designs.

【译文】一些航运公司倾向于以前成功的设计方案，几乎不允许对以往船舶设计进行改进。

【分析】本例的英译汉省略了连词 and，避免了啰唆冗余，并没有减轻对原文意思的准确表达，符合汉语的表达习惯，也体现出汉语重意合的特点。

【例 4】A floating body sinks to a depth such that it displaces its own weight of liquid.

【译文】浮体浸入液体刚好到排开与其本身重量相等的液体这样的深度。

【分析】本例省译了程度连词 such that，没有按照字面意思直译为“如此…以至于…”。特别注意，在英译汉时，要避免类似“如此…以至于…”这样的翻译腔，而是要根据原文的含义，用符合汉语表达习惯的译文通顺译出。

（3）介词的省译。大量使用介词是英语的特点之一。虽然介词不能单独作为句子成分存在，但它可以表示名词或代词等与句中其他词的关系。汉语里面介词较少，用法也没有英语介词那么复杂。在翻译时要么将介词转译为汉语里面的其他成分，要么省略不译。

【例5】The "History of Marine Animal Populations" aims to improve our understanding of ecosystem dynamics，specifically with regard to long-term changes in stock abundance，the ecological impact of large-scale harvesting by man，and the role of marine resources in the historical development of human society.

【译文】"海洋动物种群历史研究"项目旨在增强我们对生态系统发展史的理解，特别是物种多样性的长期变化，人类大范围捕捞对生态造成的影响，以及海洋资源在人类社会的历史发展中所起的作用。

【分析】本例说明了海洋动物种群历史研究项目的研究目的和主旨，"aim to"表示"旨在…"；"with regard to"表示"在…方面""关于…"。在译文中，"with regard to"采用了省译法，并不影响原文意思的表达。

【例6】The discovery of a whaling vessel wreck site，its interpretation，and understanding the historical and archaeological data in the context of our maritime past is an exciting task.

【译文】发现捕鲸船的残骸地点，阐释其缘由，理解其在人类海洋探索中的历史意义和考古价值，是一项令人兴奋的任务。

【分析】在本例中，"wreck site"是指失事的捕鲸船的残骸地点，"in the context of"表示"在…背景下"。译文中对"in the context of"并没有按照字面意思直译，而是结合上下文灵活处理。本例特别值得注意的是词类转换法。原句中三个名词词组做主语，"the discovery of a whaling vessel wreck site""its interpretation"和"understanding the historical and archaeological data in the context of our maritime past"，在译文中巧妙处理为三个动宾短语，分别对应为"发现捕鲸船的残骸地点""阐释其缘由""理解其在人类海洋探索中的历史意义和考古资料"，清晰流畅地再现了原文的含义。这是典型的词类转换法，体现了英语善用名词，汉语善用动词的特点。

（4）引导词 there 的省译。在英语中，以 there 做引导词的句子随处可见，使用范围相当广泛。there 可以与 be 动词以及其他部分不及物动词，如 seem，appear，exist，happen 等连用，构成与汉语不同的"某地有…"句式。

【例7】There's no doubt that the history of American whaling is a significant part of our national maritime heritage，for it is a topic that encompasses historic voyages and seafaring traditions set on a global stage.

【译文】毫无疑问，美国捕鲸的历史是我们国家的海洋传统中很重要的一个组成部分，因为捕鲸这个主题包含了发生在世界舞台上的各种历史航行和航海传统。

【例 8】While voyages，for example to fetch timber，are likely to have occurred for some time，there is no evidence of enduring Norse settlements on mainland North America.

【译文】即便像运送木材之类的航行可能进行了一段时间，但是没有迹象显示挪威人在北美建立了永久殖民地。

【分析】两个例子中的 there be 句型都进行了省译。在这种句式中，there 已经失去了原有的意义，翻译的时候不必强行翻译出来，句子反倒简明可读。

(5) 代词的省译。英语中的代词极多，包括人称代词、物主代词、反身代词、指示代词等，而且有一些还有词形变化，用来避免重复名词，使语言结构简洁明了。因此在翻译为汉语时，有的代词就可省略不译。

【例 9】When clothes from China arrive by ship at an entry port，the shipment has only completed part of its journey，all be it generally the longest part. It will then need to be transported from the quayside to a depot to be sent out to your local shops.

【译文】这是个漫长的旅程，当船舶从中国运输服装到达入境港口，只是完成了整个运输过程的一部分。接着需要把这些服装从码头运到仓库，最后分送到当地的商店。

【分析】本例中有两句话，第二句话中“It”指代前文中的名词短语“the shipment”，该名词短语已经在前文中翻译了出来，因此在第二句话中，则可采用省译。

(6) 副词的省译。在汉英翻译实践中，省略的更多的是副词。因为在修饰词的使用上，英汉两种语言有很大的差异。汉语为了加强语气或是使语气更为通顺，往往在名词或动词前添加一些形容词或副词。而英语在修饰词的使用上则较为慎重。因此，翻译时对原文中的修饰词要有适当的删减。凡是英语里的动词或名词已经包含了汉语里修饰名词或动词的形容词或副词的含义时，就可以把这些形容词或副词略去不译。当原文只是为了语气的通畅并无特别强调之意时或者修饰词过多时，也要适当删除一些。

【例 10】我们要积极推进海洋生态环境的保护，努力实现海洋生态平衡。

【译文】We must promote the protection of marine ecological environment to achieve its equilibrium.

【分析】像汉语“努力提高”“积极推进”“进一步加速”等这一类表达，翻译时应引起特别注意，这些副词纯属汉语加强语气的用词，在翻译中与“perfect”“improve”“promote”“accelerate”这些词连用时，切记这些词在英语中本身就含有“more”“actively”或“further”的意思。因此，原句中的“积极”“努力”二词在译文中都应去掉，选用“promote”和“achieve”，保持了译文的通畅有力。同时，“海洋生态”出现了两次，第二次直接用“its”代替。

2. 意义上需要进行的省译 和表示语法结构的虚词不同的是，实词是指具有实际独立意义并能在句子中充当句子成分的词。这些具有意义的实词组成了英语句式的主干成分，一般需要翻译出来，但是在有的情况下，也不必将全部的实词翻译出来，而需要进行适当的省译，前提是保证原文和译文的意义不会流失。

（1）重复词的省译。同义反复的词语省去不译。由于汉英两种语言的表达习惯不同，汉语中的同义反复较多，按照西方的思维是不符合逻辑的，被视为英文表达的大忌。如果译者照直把这种汉语的表达习惯译成英语，就会显得多余，不符合译文读者的语言习惯。

【例 11】海上油轮、货轮，海洋石油平台，海上输油管道，海上油田等溢油事故；船用燃油泄漏，船舶沉没；海洋石油勘探开发作业，港口、码头作业及利用海上设施、海岸设施作业过程中发生的海上溢油事故造成的生态损害，已成为威胁海洋生态安全的主要原因之一。

【译文】The ecological damage caused by marine oil spills has become one of the main reasons which threaten the marine ecological security. These oil spills include ones in the oil tanker, ship, offshore platform, floating pipe line, offshore and marine fuel leak and ship's sinking oil spill, as well as marine spills happened in the process of the offshore oil exploration and development operation, port and terminal operation and the offshore and costal facility utilization.

【分析】本例原文比较零散，并且有很多名词的排列，主要靠顿号和逗号说明句子中各成分的关系。翻译时首先要找到主句：海上溢油事故造成的生态损害已成为威胁海洋生态安全的主要原因之一。然后再分析前面的修饰部分。首先是列举了地点类的溢油事故，包括海上油轮、货轮溢油事故，海洋石油平台，海上输油管道，海上油田；然后是事件性的溢油事故，包括船用燃油泄漏，船舶沉没；最后是作业过程中的溢油事故，包括海洋石油勘探开发作业，港口、码头作业及利用海上设施、海岸设施作业过程。在译文中，用 in 加上地点，表达地点类的事故，再用 and 连接事件性的事故，最后用 as well as 连上过程中的事故。这样句子清晰，并且关系明确，避免了直译原文的罗列式杂乱。翻译时省略了很多 marine，因为整句话说的都是海上溢油事故，所以这里的省译是可以接受，并且是有必要的。

（2）重复含义的省译。

【例 12】But in more recent times，governments around the world have wised up to the fact that these gateways have the potential to bring much more into the country's economy if only they could have dedicated management and extra funds for development.

【译文】但是近年来，世界各地的政府已经意识到，如果将港口管理外包并引入额外资金，这些门户将具有很大的潜力，可以给本国带来更大的经济效益。

【分析】原文中“the fact”后跟有由“that”引导的同位语从句。在同位语的汉译中，由于句意重复，不需要译出“the fact”，只需要译出同位语从句的内容即可。

【例 13】The road freight，or haulage sector is concerned with the movement of goods by road and as a market is generally split into two groups：own transport and hire or reward.

【译文】陆运是指通过公路运输货物，可以分为两个部分：自行运输、租运或有偿运输。

【分析】在本例中，主语是“the road freight”和“haulage sector”，是通过连接词“or”连接的同义词。二者意义重复，因此在翻译时，只需要翻译出其中之一，传达出原文之意即可。

3. 修辞上需要进行的省译　在翻译中，准确流畅的译文并非是将英语中的每个词语都一五一十地翻译出来，而是要注意到译文的可接受性。英语和汉语中的修辞逻辑不仅会引起增译，也会引起必要的省译，省去那些拖沓的词或短语，可使译文简明晓畅。

【例 14】In mid-1977，world sea-going tonnage amounted to 338.5 million grt which reflected a rise of 5.8% in grt over mid-1976，compared with a corresponding increase of 9% from mid-1975 to mid-1976.

【译文】1977 年年中，世界远洋船吨位达 3.385 亿注册总吨，比 1976 年年中注册总吨增加了 5.8%，而 1976 年年中注册总吨比 1975 年年中增加了 9%。

【例 15】They float on the surface of the water and they give off light when they are disturbed.

【译文】它们在水面上漂浮，受到干扰时便放出光来。

【分析】在【例 14】中，省译了原文的"which reflects"和"compared with"；而在【例 15】中，省译了原文的"and they"，但是这样的省译丝毫不影响原文意义的再现，并且符合汉语的修辞逻辑表达习惯。

【例 16】For example，if there are language issues，the port agent will ensure that they are overcome；if there is paperwork to be filled in，the port agent will make sure it is perfect；if port services are needed for the ship call，the port agent will book them；and he/she will make sure the berth and stevedores are ready for the ship at its allotted time.

【译文】例如，港口代理会设法解决语言沟通问题；确保填报的文件准确无误；预定相关的港口服务；并确保船舶在预定时间内靠泊以及码头工人按时到位。

【分析】原文由三个包含"if"的条件状语从句组成，在处理这句话的翻译时，如果直译出条件句的内容，译文就会显得冗长复杂。译文并不直译条件句的结构"如果…就…"，而是将条件句中的主要内容提取出来。在第一个条件句中，将"language issues"与主句结合在一起，翻译成"设法解决语言沟通问题"；第二个条件句中，将"paperwork to be filled in"与主句结合在一起，翻译成"确保填报的文件准确无误"；第三个条件句中，将"port services"与主句结合在一起，翻译成"预定相关的港口服务"。此外，本例还体现了前面提到的名词的省译（the port agent）和代词的省译（he/she）。

无论是英译汉还是汉译英，省译基本都是出于译文语言习惯和语法的需要，不省去这些词，译文就显得拖泥带水，甚至会出现画蛇添足的结果。使用省译要求译者对译文有驾轻就熟的运用能力。海洋英语有自身的特点，如用语规范、用词精炼、内容确切、条理清晰、陈述客观、专业性强等。因此，海洋英语文献的翻译不同于一般英文文本的翻译，译者必须了解海洋英语的特点，掌握特定的翻译技巧。在翻译中，根据语法、语义及语境的需要，适当省去冗余成分，使译文更加流畅自然，更符合译入语的表达习惯，提高译文的准确性和专业性。

Questions for Discussion：

1. Numerate a few examples of grammatical omission.
2. Numerate a few examples of semantic omission.
3. What is the relationship between amplification and omission? How do you understand these two methods?

Ⅳ. Exercise 1: Sentences in Focus 练习 1：单句翻译

1. Translate the following sentences into Chinese

(1) What is a disaster? An event, either natural or man-made, that causes great distress or destruction or that requires a response beyond the normal capacities of the agencies involved.

(2) Remember that no plan can ever fully apply to all situations. No disaster plan should ever be cut in stone.

(3) Knowledge of the principles of marine disaster scene management along with common sense, communication skills and good seamanship will be very important if and when you encounter a multi-casualty situation.

(4) The marine environment may in general make response more difficult since an emergency at sea often involves continued hazards and complications that must be dealt with in addition to the need to care for and evacuate survivors.

(5) The Master of the distressed vessel is ultimately responsible for the welfare of his passengers and crew. This includes timely evacuation if required, developing an action plan in conjunction with SAR authorities, and sharing information on all significant developments with SAR authorities.

(6) During a marine or air incident, JRCC is the responsible authority and provides executive control of the operation. JRCC may provide an interface or liaison function with the provincial emergency program, ambulance authorities, police, any non-aligned resource or service.

(7) There is a history of cooperation between US and Canadian Coast Guards, both during exercises and on incidents. This includes the sharing of marine and air resources, as well as mutual support related to incident coordination and communications, particularly in border areas.

(8) During the incident, the coordinator gives direction to the OSC and provides a communications interface between agencies.

(9) Depending on circumstances, the OSC may delegate the actual operation of the vessel to subordinates in order to be able to concentrate on managing the incident.

(10) As soon as trained rescue personnel are on scene, the OSC or

Transport Officer will determine if the scene is STABLE or UNSTABLE, and convey this information to the triage officer and rescue personnel who may approach the scene or board the distressed vessel.

(11) If the OSC is the eyes and ears of the JRCC then the Transport Officer may be thought of as the eyes and ears of the OSC.

(12) The Triage Officer should be appointed prior to reaching the disaster scene. He or she must be familiar with the principles of marine disaster scene management and triage.

(13) The members of a Coast Guard rescue party will, after determination of scene stability and hazards, assist the Transport and Triage Officers as required.

(14) Depending on the actions of the trained rescuers, these people will either hinder or help the operation. Bystanders may be assigned jobs that are within their capability by the rescuers on scene.

(15) Uninjured survivors of a disaster often show amazing resilience and can sometimes be extremely useful in the rescue effort.

2. Translate the following sentences into English

(1) 1 000 吨原油从一艘油轮中泄漏，流入了海洋，而恶劣天气使得油轮溢油的处理工作更为复杂。

(2) 一艘雄伟的油轮可以安然驶过横流，而即便是最结实的小船，也不堪一击。

(3) 移民乘坐的船只接二连三地翻沉。上周五，又有另一艘船沉入了地中海看似平静的水中。

(4) 应急计划是指结果超出平常（预期）计划之外的计划。它经常被用于风险管理，虽然某种特殊的风险发生的可能性不大，却会带来灾难性的后果。

(5) 美国海岸警卫队是美国军队的一个分支，亦是美国七个军警部门其中一员。海岸警卫队基本职责包括海洋国土安全、海洋法实施、搜索及拯救、海洋环境保护等。

(6) 当发生搁浅、机器故障或火灾时，以下的资料及判断报告对船主是否决定雇用救难专家有所助益。

(7) 由于海上航行的特点，海难中伤亡往往十分惨重。所有船只，包括军舰，在面对天气状况、轮船设计过失或人为错误等问题时，都会变得不堪一击。

(8) 操作人员在雷雨天气工作时，有可能遭到雷击，造成意外伤亡事故。

(9) 地震引起的海啸袭击了沿海地区，导致灾难性的破坏。

(10) 医疗分类区域确定之后，指挥车带着所有需要的补给与设备就赶到了现场。

(11) 污染物直接通过城市污水和工业废水排放进入河流和海洋，有时这些废弃物有害有毒。

(12) 石油泄漏可能来自油轮、海上平台、钻机和油井中的原油泄漏，以及精炼石油产品（如汽油、柴油）及其副产品的泄漏或任何油性废弃物或废油的泄漏。

(13) 船舶会从多方面污染水道和海洋。散装货船的货物残余物排放会污染港口、航道和海洋。

(14) 按照相关国际惯例，中国应该建立并不断健全重大海难风险预警及应急预案系统。

(15) 他们把这地方变成了臭名昭著的海盗中心。海盗的利益虽然已消失，但海盗的本性不改。

Ⅴ. Exercise 2: Passages in Focus　练习 2：短文翻译

1. Translate the following passages into Chinese

Fireboat No. 7, which is an aluminium-hulled catamaran rescue boat, was put into service in 1990. It is FSD's only rescue boat designated for the purpose of mass rescue in marine areas other than the Hong Kong International Airport. Its main duties are as follows:

• to provide port safety and rescue services in Hong Kong waters, in particular to convey a large number of victims/causalities from the disaster scene at sea to a safe place or medical facilities on land in case of large-scale marine incidents such as marine fire, sinking of a large vessel, etc.;

• to provide support to fire-fighting services of other fireboats when there is a marine fire; and

• to serve as a rescue boat when nuclear-powered vessel visits Hong Kong. In case of emergency, it will be responsible for evacuating the crew members on board, monitoring their radiation level and providing them with simple decontamination facilities on the spot where necessary.

The designed life expectancy of Government aluminium-hulled vessels is around 15 years. Fireboat No. 7 has now been in service for over 20 years.

The routine annual overhaul conducted by the Marine Department has revealed that the hull and the decking plate of the fireboat are ageing notably and rusting away. The performance of the fireboat has also deteriorated with the increase in annual maintenance downtime due to mechanical fault by about 62% from 24 days in 2008 to 39 days in 2011.

In order to meet the operational requirements more effectively, we propose to procure a new vessel with the following major enhanced fire-fighting and rescue functions and installations:

- the maximum speed will increase to 35 knots from the existing 27.5 knots. A higher speed enables speedier arrival at the incident scene and conveyance of victims/casualties to a safe place or medical facilities on land;
- rescue life rafts with higher capacity (from the present 320 to 420 persons) will be equipped for mass rescue to tie in with the commissioning of the cruise terminal 2. In the event of a cruise incident, casualty of over a thousand might be involved;
- a small boat will be provided to facilitate fire services staff conducting incident scene assessment at shallow waters and formulating action plan;
- a high-efficient air filtration system and radiation monitoring equipment etc. will be installed and the wheelhouse/cabin will have pressurization systems. These are to prepare for rescue operations just in case there is such a need during the visits of nuclear-powered vessels to Hong Kong. The new vessel will also be better equipped with decontamination facilities to enhance our capability in dealing with relevant incidents and provide better protection for the frontline staff.

2. Translate the following passages into English

"医疗类选法"的定义是：根据轻重缓急，对病人，尤其是战争中和灾难中的受害者进行分类治疗，以最大限度地提高幸存者的比例。如果灾难现场有足够多的救援人员和资源，那么每位伤者都将得到同样努力的治疗，就像伤者在单一伤亡事件中所得到的治疗那样。不幸的是，在海难中常常不是这种情况。想想看，有些人不管得到什么样的治疗，都会生存下来，而有些人不管救援如何努力，也必然死亡。医疗类选法的主要目标是确定哪些人可以通过早期干预和治疗而存活下来。如果我们能做到这一点，并提供可以挽救生命的干预措施，那么我们就能最大限度地提高幸存者的比例。

救援单位应该在其职责区域内预先做好计划，以应对高风险或重大影响的事件。这些事件可能包括主要的客运交通或商运交通，沿海机场或与海上通道相连接的其他基础设施，或特定的高风险活动。要考虑到可能的危险，潜在的伤亡人数，进口和出口，以及可能的集结待命区域。还要考虑到将对事件做出应对的其他机构或其他资源，考虑到他们的责任和能力，以及你们将如何协力工作。

从接收到海难事件的第一次通知开始，就应该考虑如何进行医疗类选。即使在路上，医疗类选人员也必须对总体形势的严重性有全面的考量。最好的办法是考虑该事件的发生过程（例如，撞击的速度、火势的严重程度、船体浸入水中的长度等），以及可能的伤亡数目。基于到达海难现场之前掌握的这些信息，救援人员随之可以开始制订医疗类选计划。然而，当他们到达现场之后，还必须根据实际需要对计划再做调整。

有一种说法是，整场救援事件的成败在于到达现场的最初 10 分钟，因为所有最重要的决策都是在这 10 分钟内做出的。确认了海难现场是否稳定之后，要完成的第一个任务就是医疗类选。由于伤亡情况可能会发生改变，那么医疗类选必须是一个持续的过程，在事件的整个过程中不断重估伤亡情况。

Unit Six　Marine Protection

第六单元　海洋保护

Ⅰ. Warm-up for Theme-related Words/Expressions 词汇预习

accountable　*a.* 负有责任的
advisable　*a.* 明智的；可取的；适当的
allocation　*n.* 配给，分配；分配额
aquaculture　*n.* 水产养殖；水产业
auditor　*n.* 审计员；查账员
autonomy　*n.* 自主权；自治，自治权
Baltic Sea　（欧洲）波罗的海
baseline　*n.* 基准线，零位线
Bornholm　*n.* （丹麦）博恩霍尔姆岛
bureaucracy　*n.* 官僚主义；官僚机构
classify　*v.* 分类，归类；把…列为密件
constancy　*n.* 稳定性；持续性；持久性
contiguous　*a.* 邻近的，共同的
cornerstone　*n.* 奠基石；基础
diagram　*n.* 图表
dilemma　*n.* 窘境，困境
disjoint　*v.* （使）脱节，（使）解体
dominant　*a.* 占优势的；统治的
encompass　*v.* 围绕，包围
equitable　*a.* 合理；公正的
federal　*a.* 联邦（制）的，同盟的
framework　*n.* 框架，构架；（体系的）结构
geographical　*a.* 地理学的，地理的
go into effect　生效
Great Lakes　（美国）五大湖区
hub　*n.* 中心，焦点
indigenous　*a.* 土生土长的；本地的
intertidal zone　潮间带
inventory　*n.* 目录，清单
legislation　*n.* 立法，制定法律
likewise　*adv.* 同样地；也，而且
marine management　海洋管理
marine protected areas（MPAs）海洋保护区
marine ecosystem　海洋生态系统
Mesoamerican　*n.* 中美洲
National Marine Sanctuaries　国家海洋保护区
NOAA　〈美〉国家海洋和大气局
NOS（The National Ocean Service）美国国家海洋局

observation *n.* 观察；观察力
participate *v.* 参加某事；分享某事
permanence *n.* 永久，持久
pertinent *a.* 相关的
preference *n.* 偏爱；优先权
privileged *a.* 享有特权的；特许的
prominent *a.* 著名的；突出的
protocol *n.* 协议
quarter *n.* 一年中的某一季
ratify *v.* 批准，许可
sanction *n.* 约束；制裁；批准 *v.* 批准；鼓励
Scandinavia *n.* （欧洲）斯堪的纳维亚（半岛）
scatter *v.* 分散；撒开
seamount *n.* 海山；海峰
sector *n.* 部门；领域
span *v.* 持续，包括
stewardship *n.* 管理工作（侧重于管理的职责，而且特指财产方面）
supranational *a.* 超国家的，超民族的
tribal *a.* 部落的，部族的
uncover *v.* 揭露，发现
underlie *v.* 构成…的基础（或起因）
UNEP（United Nations Environment Programme） 联合国环境规划署

Ⅱ. Lead-in & Reading 课文导读

Marine Protected Areas

(1) Marine protected areas (MPAs)[1] are areas of the oceans or Great Lakes that are protected for a conservation purpose. If you have ever gone fishing in central California, diving in the Florida Keys, or boating in

Thunder Bay, you have visited one of these MPAs. In the United States, there are over 1,600 MPAs spanning a range of habitats, including the open ocean, coastal areas, inter-tidal zones[2], estuaries, and the Great Lakes. Nearly all of these areas allow multiple uses. About 41 percent of U. S. marine waters are protected in some way, with three percent in highly protected no-take MPAs to protect sensitive species and habitats.

National System of MPAs

(2) With different federal, state, tribal and local agencies managing the more than hundreds of marine protected areas located all over the U. S., what is going on in each can get a little disjointed and opportunities to coordinate or share lessons learned can be missed. So wouldn't it be nice if there were some system to coordinate planning and management of our nation's MPAs? Guess what? There is!

(3) Managed by the federal government, the national system of MPAs brings work together at the regional and national levels to achieve common objectives for conserving the nation's important natural and cultural resources. The MPAs in the system are still managed independently, but they now have a framework to tie them all together. There are currently 437 members of the national system of MPAs. Over time, the National Marine Protected Areas Center will continue to work with existing U. S. MPAs to connect, strengthen, and expand the nation's MPA programs.

Restrictions

(4) One question many people have about marine protected areas is whether or not there are restrictions associated with the use of these areas. The answer is that...it depends. MPAs are established for the conservation of their natural or cultural resources. While there can be restrictions on certain activities in MPAs, nearly all U. S. MPAs allow multiple uses, including fishing. Some areas, such as marine reserves, are more restrictive, limiting the catching of fish, collection of shells, or other activities where something may be removed from the area. Marine reserves are sometimes referred to as "no take" MPAs, and occupy about three percent of U. S. waters.

Classifying MPAs

(5) Because marine protected areas vary widely, the National MPA Center developed a system to help describe these areas using characteristics that are common to most MPAs. The characteristics include conservation focus, level of protection, permanence of protection, constancy of protection, and ecological scale of protection. The end result is a common vocabulary for MPA managers, something that comes in handy when exchanging ideas and lessons learned or working to identify additional areas that should be protected.

MPA Inventory

(6) Interested in finding out if there are marine protected areas where you live or have visited? Check out the Marine Protected Areas Inventory. This online tool lets you use an interactive map to view the MPA Inventory sites and associated data, query sites by specific conservation attributes, or to search and view sites by region. Managers can use the Inventory for marine management and conservation planning. In fact, the primary purpose of the Inventory is to maintain baseline information on MPAs to the assist in the development of the National System of MPAs.

NOAA's National MPA Center

(7) The National MPA Center was established in 2000 following Executive Order 13158. The executive order was issued to help protect the significant natural and cultural resources within the marine environment for the benefit of present and future generations. The order directs the Department of Commerce, the Department of the Interior, and other federal agencies to work closely with states, territories, tribes, fishery management councils, and groups with an interest in marine resource conservation to develop a scientifically-based, comprehensive National System of MPAs representing diverse U. S. marine ecosystems.

(8) The MPA Center is located within NOAA's National Ocean Service and is a division of the Office of National Marine Sanctuaries.

Science and Stewardship

(9) The MPA Center uses science to assess the nature of MPAs and how they are used to sustain healthy marine ecosystems. The MPA Center focuses its objectives on enhancing MPA stewardship by strengthening capacity for planning, management, and evaluation. For example, the MPA Center is working with other MPA programs to share lessons about how they are planning for and adapting to climate change.

MPA Programs

(10) There are many different types of MPAs including national marine sanctuaries and national estuarine research reserves.

(11) The 13 national marine sanctuaries, managed directly by NOS, and the 28 national estuarine research reserves, managed by states in partnership with NOS, are conserved areas that are also hubs for recreation, research, and education. Other types of MPAs include national parks, national wildlife refuges, and state areas for the protection of habitat, fish, and wildlife. The National System of MPAs is helping to weave all of these areas together into an effective network that can protect species that move through various habitats during different life stages, and that can meet common challenges faced by MPA programs.

National Efforts

(12) The marine protected area network is still in its infancy. As of October 2010, approximately 6,800 MPAs had been established, covering 1.17% of global ocean area. Protected areas covered 2.86% of exclusive economic zones (EEZs)[3]. MPAs covered 6.3% of territorial seas. Many prohibit the use of harmful fishing techniques yet only 0.01% of the ocean's area is designated as a "no take zone[4]". This coverage is far below the projected goal of 20%-30%. Those targets have been questioned mainly due to the cost of managing protected areas and the conflict that protections have generated with human demand for marine goods and services.

Greater Caribbean

(13) The Caribbean region; the UNEP-defined region also includes the

Gulf of Mexico. This region is encompassed by the Mesoamerican Barrier Reef System proposal, and the Caribbean challenge. The Gulf of Mexico region (in 3D) is encompassed by the "Islands in the Stream" proposal.

(14) The Greater Caribbean subdivision encompasses an area of about 5,700,000 square kilometres (2,200,000 sq mi) of ocean and 38 nations. The area includes island countries like the Bahamas and Cuba, and the majority of Central America. The Convention for Protection and Development of the Marine Environment of the Wider Caribbean Region (better known as the Cartagena Convention) was established in 1983. Protocols involving protected areas were ratified in 1990. As of 2008, the region hosted about 500 MPAs. Coral reefs are the best represented.

South Pacific

(15) The South Pacific network ranges from Belize to Chile. Governments in the region adopted the Lima convention and action plan in 1981. An MPA-specific protocol was ratified in 1989. The permanent commission on the exploitation and conservation on the marine resources of the South Pacific promotes the exchange of studies and information among participants.

(16) The region is currently running one comprehensive cross-national program, the Tropical Eastern Pacific Marine Corridor Network, signed in April 2004. The network covers about (211,000,000) square kilometres (81,000,000 sq mi). One alternative to imposing MPAs on an indigenous population is through the use of Indigenous Protected Areas, such as those in Australia.

North Pacific

(17) The North Pacific network covers the western coasts of Mexico, Canada, and the U. S. The "Antigua Convention" and an action plan for the North Pacific region were adapted in 2002. Participant nations manage their own national systems. In 2010 - 2011, the State of California completed hearings and actions via the state Department of Fish and Game to establish new MPAs.

United States and Pacific Island Territories

(18) President Barack Obama signed a proclamation on September 25, 2014, designating the world's largest marine reserve. The proclamation expanded the existing Pacific Remote Islands Marine National Monument, one of the world's most pristine tropical marine environments, to six times its current size, encompassing 490,000 square miles (1,300,000 km^2) of protected area around these islands. Expanding the monument protected the area's unique deep coral reefs and seamounts.

(19) Diagram illustrating the orientation of the 3 marine sanctuaries of Central California: Cordell Bank, Gulf of the Farallones, and Monterey Bay. Davidson Seamount, part of the Monterey Bay sanctuary, is indicated at bottom-right. In April 2009, the US established a United States National System of Marine Protected Areas, which strengthens the protection of US ocean, coastal and Great Lakes resources. These large-scale MPAs should balance "the interests of conservationists, fishers, and the public". As of 2009, 225 MPAs participated in the national system. Sites work together toward common national and regional conservation goals and priorities. NOAA's national marine protected areas center maintains a comprehensive inventory [40] of all 1,600+ MPAs within the US exclusive economic zone. Most US MPAs allow some type of extractive use. Fewer than 1% of U. S. waters prohibit all extractive activities.

(20) In 1981, Olympic National Park became a marine protected area. The total protected site area is 3,697 square kilometres (1,427 sq mi). 173.2 km^2 of the area was an MPA. The national system is a mechanism to foster MPA collaboration. Sites that meet pertinent criteria are eligible to join the national system.

(21) In 1999, California adopted the Marine Life Protection Act, establishing the first state law requiring a comprehensive, science-based MPA network. The state created the Marine Life Protection Act Initiative. The MLPA Blue Ribbon Task Force and stakeholder and scientific advisory groups ensure that the process uses the science and public participation.

(22) The MLPA Initiative established a plan to create California's statewide MPA network by 2011 in several steps. The Central Coast step was

successfully completed in September, 2007. The North Central Coast step was completed in 2010. The South Coast and North Coast steps were expected to go into effect in 2012.

United Kingdom and British Overseas Territories

(23) The United Kingdom is responsible for 6.8 million square kilometres of ocean around the world, larger than all but four other countries. There are a number of marine protected areas around the coastline of the United Kingdom. The UK is also creating marine protected reserves around several British Overseas Territories.

(24) In March 2015, the UK announced the creation of a marine reserve around the Pitcairn Islands in the Southern Pacific Ocean to protect its special biodiversity. The area of 830,000 square kilometers (322,000 square miles) will make it the world's largest contiguous marine reserve.

(25) In January 2016, the UK government announced the intention to create a marine protected area around Ascension Island. The protected area will be 234,291 square kilometers, half of which will be closed to fishing.

Europe

(26) The Natura 2000 ecological MPA network in the European Union included MPAs in the North Atlantic, the Mediterranean Sea and the Baltic Sea. The member states had to define NATURA 2000 areas at sea in their Exclusive Economic Zone.

(27) Two assessments, conducted thirty years apart, of three Mediterranean MPAs, demonstrate that proper protection allows commercially valuable and slow-growing red coral (Corallium rubrum) to produce large colonies in shallow water of less than 50 metres (160 ft). Shallow-water colonies outside these decades-old MPAs are typically very small. The MPAs are Banyuls, Carry-le-Rouet and Scandola, off the island of Corsica.

Notes:

[1] marine protected areas (MPAs) 海洋保护区。国际自然保护联盟（IUCN）将海洋保护区定义为“任何通过法律程序或其他有效方式建立的，对其中部分或全部环境进行封闭保护的潮间带或潮下带陆架区域，包括其上覆水体及相关的动植物群落、历史及文化属性”（IUCN，1994）。该定义涵盖内容相对广泛，凡是符合国际自然保护联盟保护区目标

的各种类型和规模的海洋保护区都包括在内。

[2] inter-tidal zones 潮间带，是指大潮期的最高潮位和大潮期的最低潮位间的海岸，也就是海水涨至最高时所淹没的地方至潮水退到最低时露出水面的地方。在潮间带以上，海浪的水滴可以达到的海岸，称为潮上带。在潮间带以下，向海延伸至约 30 米深的地带，称为亚潮带。

[3] exclusive economic zones（EEZs） 专属经济区，又称经济海域，是指国际公法中为解决国家或地区之间的领海争端而提出的一个区域概念。专属经济区是指领海以外并邻接领海的一个区域，专属经济区从测算领海宽度的基线量起，不应超过 200 海里（370.4 千米）。

[4] no take zone 禁渔区，是指全面禁止一切捕捞生产或禁止部分作业方式进行捕捞的水域，为保护某些重要的经济鱼类、虾蟹类或其他水生经济动植物资源，在其产卵繁殖、幼鱼生长、索饵育肥和越冬洄游期所划定的禁止或限制捕捞活动的水域，如印度洋禁捕区禁止商业性捕鲸。

Ⅲ. Teaching Focus (6): Affirmation and Negation in Translation 翻译技巧 (6): 正译与反译

汉英两种语言在表达正说和反说时有很大差异，尤其英语在否定意义的表达上更为复杂，有时形式否定而实质肯定，有时形式肯定而实质否定。在两种语言互译时，原文中正说的句子可能不得不处理成反说，或是用反说表达更为合适。反之亦然。在翻译中，这种把正说处理成反说、把反说处理成正说的译法，称为正、反译法。

笼统地说，英语句子中含有“never”“no”“not”“un-”“im-”“in-”“ir-”“-less”等否定词以及否定前缀或后缀的单词，以及汉语句子中含有“不”“没”“无”“未”“甭”“别”“休”“莫”“非”“勿”“毋”等否定词的即为反说，不含有这些否定词的即为正说。但在实际操作时，正说和反说的界限又变得极为模糊，例如“correct”可以翻译成“正确”（正说），也可以翻译成“没有毛病”（反说）。因此，到底译文要采用正说还是反说，完全要看译文语言的惯用表达和上下文的语气语态。“正说反译”和“反说正译”大致可分为以下几种情况：

1. 英译汉正说反译 英语中有些否定概念是通过含有否定意义或近似否定意义的词来表达的，虽然形式是肯定的，但这类词大多是某些肯定词所引申或变化出来的反义词，或经过长期历史演变而引申出其他否定词义，即所谓的“含蓄否定词”或“暗指否定词”，这类词在译成汉语时，需要变成汉语的否定词组，必要时还需要做词类转换。

（1）名词。含蓄否定名词主要有：shortness / shortage（不够；不足）、lack（缺乏；没有）、absence（不在）、failure（未能；不成功）、defiance（不顾；无视）、denial（否认；否定）、exclusion（排除）、freedom（不；免除）、refusal（不愿；不允许）、loss（失去）等。

【例 1】Lack of experience，the country's new long-term energy policy includes hydropower but puts more emphasis on solar and wind，complemented by geothermal，biomass and ocean energy.

【译文】由于经验不足，该国新的长期能源政策将氢能纳入其中，但重点则放在了太阳能和风能上，同时兼顾地热能、生物质能和海洋能源。

【分析】"lack"的意思是"缺少""短缺"，用"正说反译"的方法来说就是"不足"。本例在翻译技巧上无太多可言，需注意"geothermal（地热能）"和" biomass（生物质能）"等专有名词的译法。

【例 2】Sea denial is a military term describing attempts to deny the enemy's ability to use the sea without necessarily attempting to control the sea for its own use.

【译文】海上拒止（或译为海上阻绝、海上阻遏），是军事术语，指试图阻止敌对方的海上力量，但不一定企图控制该海域。

（2）形容词或形容词词组。含蓄否定形容词主要有：absent（不在；不到；不出席）、blind（看不到；不注意）、awkward（不熟练；不灵活）、bad（令人不愉快的；不舒服的）、dead（无生命的；无感觉的；不毛的）、difficult（不容易的）、foreign to（不适合的；与…无关）、short of（不够）、poor（不好的；不幸的；不足的）、thin（站不住脚；不够有力）等。

【例 3】He was deaf to the advice that he should return to Seattle after high school，working as a fisherman and later as maritime-industry metal worker.

【译文】人们建议他高中毕业后回到西雅图去当一个渔夫，然后成为一名海洋工业的金属工人，但他对此置若罔闻。

【分析】译句将"deaf"处理为"置若罔闻"，表示"不听""听不进"的意思，采用的是正说反译的策略。由于从句的内容过长，此处将主句和从句按意群断开，主句独立成句，用"此"字连接前后文，使得译句更为通顺，避免拖沓。

（3）动词或动词短语。英语中常见的含蓄否定动词包括：refuse（不愿；不肯；无法）、lack（缺乏；没有）、defy（不服从；不遵守；不让）、forbid（不许）、stop（不准；别）、ignore（不理；不肯考虑；无视；不顾）、hate

(不愿意)、miss(没听清楚;没赶上)等。

【例4】Wrecks have been largely overlooked by marine archaeology until now—much like the little-known history of black scuba divers.

【译文】如同黑色潜水员那段鲜为人知的历史一样,沉船残骸直到现在才得到了海洋考古学的重视。

【分析】本例中没有按"overlook"的本义把前半句译为"在很大程度上被海洋考古所忽略",而是倒过来说"直到现在才得到了海洋考古学的重视",这样一来就非常通顺易懂了。在句子处理的层面上,译文颠倒了原文的语序,根据时间逻辑先译出后半句的事实背景,然后才交代前半句中的判断表态,照顾了两种思维的差异。

【例5】Scientists denied the rumor that a volcano in the South Atlantic Ocean might have seriously harmed the world's largest chinstrap penguin colony.

【译文】有谣言称,坐落于南大西洋的一座火山已经严重破坏了世界上最大的帽带企鹅栖息地。科学家已澄清了这一谣言。

【分析】本例中的"deny"带有否定含义,可译为"否认",而译句译为了"澄清"。由于从句部分过长,译者根据语义及时断句,并根据汉语习惯,将表示判断表态的"deny"部分置于句末。

(4)副词。

【例6】Presumably, since you work in a field that deals with the ocean, boats, sailors and the navy, you may be a marine meteorologist.

【译文】不出所料,既然你从事的领域与海洋、船只、水手或者海军有关,那你就可能是个海洋气象学家。

【分析】译文将没有否定含义的"presumably"转换成了带有"不"字的"不出所料",而"work in a field"也从"动词+介词+名词"的结构转换为了"修饰语加名词"的结构。

【例7】The scientist slowly showed evidence to convince his opponents that ocean acidification has damaged certain marine organisms such as oysters and corals.

【译文】这位科学家不慌不忙地拿出了证据,他想要让反对者相信,海洋酸化已经破坏了诸如牡蛎和珊瑚等的海洋生物。

【分析】"slowly"处理为"不慌不忙",体现了"科学家"自信从容的态度。另外,译句按照意群将原句切成了3个小分句,充分考虑到了汉语的行文习惯——汉语多短句,英语多长句,照顾了两种语言的差异。

(5) 介词和介词短语。含蓄否定介词和介词短语主要有：without（没有）、against（不符合）、except（不包括）、beyond（无；无法）、beside（不对；不符合）、above（不至于）、behind（未能；尚未；还没有）、instead of（没有；而不是）、aside from（不包括）、far from [绝不是；还不；非但不…(甚至) 还…]、at large（未被捕；未捉拿归案）、at a loss（不知所措；不知如何是好；完全不明白/不理解）、but for（要不是/没有…就不能/不会/无法…）、would rather...than（宁愿/宁可…也不）等。

【例 8】Many prohibit the use of harmful fishing techniques yet only 0.01% of the ocean's area is designated as a "no take zone". This coverage is far below the projected goal of 20%-30%.

【译文】许多水域都禁止使用有害的捕捞工艺捕捉鱼类，但只有 0.01%的海洋区域是"禁捕区"。此覆盖率远达不到预期 20%～30%的目标。

【分析】"be far below" 也可以正译为"远低于"，但此处刻意采用了反译"远达不到"，用以锻炼逆向思维的能力。此外，本句中的"of"一词在翻译时也应注意，"of"前后有讲究，其后一般是具体微观的内容，前边是宏观概括的内容，翻译时应当根据汉语习惯调整语序。"coverage"翻译成了"覆盖率"采用了增词的技巧，在"覆盖"的原意之后添上了"率"，使之更符合语境。

【例 9】The minister said that the two officials discussed maritime issues with dignity, stressing the importance of not letting disputes escalate.

【译文】部长说，两位官员不失尊严地探讨了海洋方面的问题，并强调不让争端升级是一件非常重要的事情。

【分析】本例将"with dignity"（体面地）正说反译为"不失尊严地"，使译文的阅读效果得以明显提升。如果按字面处理成"两位官员体面地探讨了海洋方面的问题"，读上去显得非常怪异，也不符合中文的行文习惯。后半句的"stressing the importance of not letting disputes escalate"也未处理成"强调不让争端升级的重要性"，而是处理成了一句由主系表构成的句子，体现了"词性转换、句式转换"的翻译技巧。

【例 10】Another dog lives at ease on the same marine base, the spokesman said, and military officials "will follow up with supportive care for it".

【译文】发言人说道，另外有一条狗在同一处海洋基地中无拘无束地生活着，军方官员也表示，"将会继续为这条狗提供后续的关怀"。

【分析】"at ease" 指的是"处于放松的状态"，可译为"无拘无束"。本例中将“the spokesman said”（发言人说到）部分提至句首，对汉语的语序做出了调整，使之更符合中文的表达习惯。

2. 英译汉反说正译　英语中明显使用not、no等否定词或否定前后缀的反说语句有时却含有肯定的含义，在这种情况下，在译成汉语时，可以用肯定形式来表达。

（1）某些带有否定含义的名词，如nothing，dislike，disbelief，unemployment，impatience等。

【例11】This spring，marine biologists flew a drone over the Sea of Cortez to capture samples of the fluids prayed from the blowholes of blue whales. However，they returned with nothing in hands.

【译文】今年春天，一些海洋生物学家驾驶一架无人机飞跃科尔特斯海，想要捕获蓝鲸气孔喷出的流体样品。然而他们空手而归。

【分析】原句里的“nothing”可以按原意处理为“他们什么也没有得到”，但如果用上“反说正译”的技巧处理成“他们空手而归”，阅读效果显得更好。因为汉语的行文是习惯使用“四六字”的结构，比如“高高兴兴”“美轮美奂”“高楼大厦”等。

（2）某些带有否定含义的形容词，如unhappy，impatient，illiterate，unusual，untouchable，careless，uneasy，uncommon等。

【例12】She gave an indefinite answer to how she has learned to read the tides，work a marine radio and navigate with a compass.

【译文】关于她是怎么学会看浪潮，怎么学会使用航海无线电台，又怎么学会利用指南针进行航海定位的，她只给了一个很模糊的回答。

【分析】“indefinite”有一个表示否定的前缀“in-”，因此不少翻译初学者会将其译为“不清楚的”“不明确的”，但转换一下思维，我们可以将它处理成“模糊的”。译文在how引导的从句中每个动词前都加了“怎么学会”，让英语中省略的信息在译文中一一呈现，体现了汉语文体重重复而英语文体重省略的行文风格。

（3）某些词组和句型，如no more than（仅仅），in no time（立刻），not more than（最多），none but（只有），none other than（正是），cannot help（情不自禁），not until（直到）等。

【例13】Marine life is everything，with no exception，that lives in the water：plants，fish，sharks，crabs，seahorses，giant squid，and the list goes on and on.

【译文】海洋生物就是一切生长在水里的东西：植物、鱼类、鲨鱼、螃蟹、海马、巨型乌贼，还有很多其他东西。

【分析】“with no exception”可以按字面译为“无一例外”，也可以反说正译为“一切”，具体的选择应由译者和语境做决定。而原句中的“everything，with no exception”是语义上的重复，为的是强调“一切”，译为汉语时则只需译一次即可，不必重复。

【例 14】Examples include none other than addressing air pollution, increasing the use of low-carbon energy in chemical engineering and deep-sea stations for ocean exploration.

【译文】范例包括的正是解决空气污染问题，加大化学工程中低碳能源的使用率，以及增加海洋探究过程中的深海站数量。

【分析】译文除了“反说正译”，将“none other than”处理成“正是”以外，还采用了“增词”的翻译技巧，例如在“空气污染”之后添加“问题”，在“使用”之后添加“率”等，使之更符合中文的语言习惯。这说明翻译时不应拘泥于某一种特定的格式，而应灵活自然地使用各类翻译技巧。

Questions for Discussion:

1. Numerate a few words that belong to the implied negation.
2. How do you understand affirmation and negation? Is there anything common between the two concepts?

Ⅳ. Exercise 1: Sentences in Focus　练习 1：单句翻译

1. Translate the following sentences into Chinese

(1) With different federal, state, tribal and local agencies managing the more than hundreds of marine protected areas (MPAs) located all over the U. S., what is going on in each can get a little disjointed and opportunities to coordinate or share lessons learned can be missed.

(2) Over time, the National Marine Protected Areas Center will continue to work with existing U. S. MPAs to connect, strengthen, and expand the nation's MPA programs.

(3) Some areas, such as marine reserves, are more restrictive, limiting the catching of fish, collection of shells, or other activities where something may be removed from the area.

(4) The characteristics include conservation focus, level of protection, permanence of protection, constancy of protection, and ecological scale of protection.

(5) This online tool lets you use an interactive map to view the MPA Inventory sites and associated data, query sites by specific conservation attributes, or to search and view sites by region.

(6) The order directs the Department of Commerce, the Department of the Interior, and other federal agencies to work closely with states, territories, tribes, fishery management councils, and groups with an interest in marine resource conservation to develop a scientifically-based, comprehensive National System of MPAs representing diverse U. S. marine ecosystems.

(7) The MPA Center focuses its objectives on enhancing MPA stewardship by strengthening capacity for planning, management, and evaluation.

(8) The 13 national marine sanctuaries, managed directly by NOS, and the 28 national estuarine research reserves, managed by states in partnership with NOS, are conserved areas that are also hubs for recreation, research, and education.

(9) Those targets have been questioned mainly due to the cost of managing protected areas and the conflict that protections have generated with human demand for marine goods and services.

(10) The permanent commission on the exploitation and conservation on the marine resources of the South Pacific promotes the exchange of studies and information among participants.

(11) One alternative to imposing MPAs on an indigenous population is through the use of Indigenous Protected Areas, such as those in Australia.

(12) In April 2009, the US established a United States National System of Marine Protected Areas, which strengthens the protection of US ocean, coastal and Great Lakes resources.

(13) The UK is also creating marine protected reserves around several British Overseas Territories.

(14) The protected area will be 234,291 square kilometers, half of which will be closed to fishing.

(15) Two assessments, conducted thirty years apart, of three Mediterranean MPAs, demonstrate that proper protection allows commercially valuable and slow-growing red coral (Corallium rubrum) to produce large colonies in shallow water of less than 50 metres (160 ft).

2. Translate the following sentences into English

(1) 在南大西洋，建立了深水海洋保护区，以保护深水鱼类物种及其栖息地不受捕捞影响。

(2) 各国应尽力积极参加区域性和全球性计划，以获取相关知识用于评估污染的性质和程度，污染范围，以及污染产生的途径、危险和补救办法。

(3) 各国在落实欧盟条例时有自治权，但当国家层面决策出现分歧的时候，最后被问责的是欧盟的官僚机制，而非国家政府部门。

(4) 沿海国在其领海内行使主权，可制定法律和规章，以防止、减少和控制外国船只，包括行使无害通过权的船只对海洋的污染。

(5) 这一方案还将与南太平洋应用地球科学委员会在数据抢救和发展深海矿物资源数据库以及（专属经济区和扩展大陆架）海洋边界方面的现行活动挂钩。

(6) 禁捕鱼区对渔业实施限制，可能会促进其他海水区域的生产力，因为可供海豚捕食的猎物会增加，而且亦能减少海床所受到的滋扰。

(7) 渔业委员会在 2011 年 2 月的会议上审议了有关保护生物多样性的具体活动，包括建立海洋保护区和海洋保护区网络，以及开展影响评估。

(8) 不少区域渔业管理组织已采取海域关闭和其他基于海域措施，以应对捕捞的影响。国际大西洋金枪鱼保护委员会数次采用时段/海域关闭措施，主要是为了保护蓝鳍金枪鱼、剑鱼和大眼金枪鱼等金枪鱼品种的幼鱼。

(9) 联合国环境规划署与合作伙伴一起，正在执行和实施全球环境基金的跨界水域评估方案，其目的是为跨界水域系统的现状和不断变化的状况制定评估方法。

(10) 中国应该在沿海人口密集与经济发达地区建立一体化的海洋生态环境研究和监测体系，包括环境监测设施、研究机构、实验室、户外观测点和生态修复示范工程。

(11) 评估对于更好地了解海洋生态系统的现状、趋势和状况十分重要。评估尤其有助于衡量各种生态系统的脆弱性、复原力和适应能力。

(12) 污染的类型包括：陆地来源的污染；海底活动造成的污染；来自“区域”内活动的污染；倾倒造成的污染；来自船只的污染；以及来自大气层

或通过大气层的污染。

（13）政府间海洋学委员会还举办了关于数据和信息管理、海平面数据分析、建模、海洋生物多样性和遥感在沿海管理中的应用等培训课程。

（14）关于在深海和公海的生物多样性问题，国际自然资源保护联盟参与了全球海洋生物多样性倡议。

（15）全球海洋观测系统是一个永久的全球性的海洋变量观测、建模和分析系统，以支持全球海洋业务。

Ⅴ. Exercise 2: Passages in Focus 练习 2：短文翻译

1. Translate the following passages into Chinese

Available information shows that there has been a significant increase in coverage of protected areas over the past decade. However, many ecological regions, particularly in marine ecosystems, remain underprotected, and the management effectiveness of protected areas remains variable. Of 232 marine eco-regions, 18 percent meet the target for protected area coverage of at least 10 percent, while half have less than 1 percent protection. The total number of marine protected areas now stands at approximately 5,880, covering over 4.7 million square kilometres, or 1.31 percent of the world's ocean area. The total global marine protected area coverage is largely composed of a relatively small number of very large marine protected areas, almost all of which are within national jurisdiction.

A recent report highlighted some of the costs and benefits of marine protected areas. While the costs of implementation, maintenance and adaptive management can be high, data on the costs of creation and management of marine protected areas and area networks remains limited. In 2002 estimates of the annual cost of running individual areas ranged from $9,000 to $6 million. In 2004 estimates put at $5 billion to $19 billion the cost of a global network that met 20 to 30 percent of protection goals. Among the benefits, the report outlined benefits for fisheries, tourism, spiritual, cultural, historical and aesthetic values, disaster mitigation, research, education and stewardship for ocean awareness and protection.

In September 2010 and with effect from 12 April 2011, the parties to the Convention for the Protection of the Marine Environment of the North-East

Atlantic (OSPAR Convention) agreed to designate six high-seas marine protected areas: Milne Seamount Complex; Charlie Gibbs South; Altair Seamount High Seas; Antialtair Seamount High Seas; Josephine Seamount High Seas; and Mid-Atlantic Ridge North of the Azores High Seas. When combined with the network of sites within national jurisdiction, these marine protected areas provide coverage of 3.1 percent of the total OSPAR Convention area. Some of those marine protected areas overlie the outer continental shelf of a coastal State. While the Charlie Gibbs South and Milne Seamount Complex areas aim at protecting and conserving the biodiversity and ecosystems of the seabed and superjacent waters, the other four areas were established to protect and conserve the biodiversity and ecosystems of the water superjacent to the sites, in coordination with, and complementary to, protective measures taken by Portugal for the seabed.

2. Translate the following passages into English

海洋保护区和保护区网络，作为更广泛的沿海管理和海洋管理框架的一部分，被认为是一项通过保护重要的栖息地，从而帮助生态系统保持健康和发挥生态功能的重要工具。然而，海洋保护区要实现其目标，就需要进行有效的设计和管理，同时考虑到利益相关者的社会经济需要。它们还必须是在所有方面进行有效管理的更大框架的一部分，并与其他工具协同行动。

生物多样性公约第十次缔约方会议通过了新的战略计划，以便到2020年大幅度减少生物多样性的丧失。计划的二十多项目标中，有若干项涉及海洋生物多样性，包括国家管辖范围外的海域。特别是会议还商定，到2020年时至少有10%的沿海和海洋地区，尤其是对生物多样性和生态系统服务特别重要的海域，通过开展有效和公平管理、建立具有生态代表性和连贯畅通的保护区系统以及其他有效的海域保护措施，使其得到养护，并整合成宽阔的海洋景观。

海洋学委员会秘书处在提供的资料中指出，国家管辖范围外的海洋保护区网络或在此海域内采取的任何其他管理行动，都要有一个监测系统和一个用于制定政策的强有力的循证依据。经常进行可靠的观察很重要，因为海洋特征在不断变化。在这方面，边界固定的海洋保护区不一定能够提供保护浮游生物多样性所需要的保护，因此，海洋学委员会正沿用电子海图的例子，探讨使用海洋保护区动态边界解决问题。委员会还注意到，国家管辖范围外海洋保护区的执法工作取决于是否有船舶跟踪系统和遥感手段。海洋学委员会已经与欧洲科学基金会海洋委员会一起设立了一个工作组，提供一个框架，以便就今后的海洋保护区规划为利益相关者提供信息，促使其参加并增强其能力。工作组正在审查和总结设立和建立海洋保护区应该考虑的各种因素；审查对已设立的海洋保护区进行评估的标准；制订一个评价海洋保护区效率和绩效标准的清单。

Unit Seven　Island Tourism

第七单元　海岛旅游

Ⅰ. Warm-up for Theme-related Words/Expressions　词汇预习

advent　*n.* 到来，出现
allure　*n.* 诱惑
Antigua　*n.* 安提瓜岛
Bahamas　*n.* 巴哈马群岛
Barbados　*n.* 巴巴多斯（拉丁美洲国家）
Barbuda　*n.* 巴布达岛
bauxite　*n.* 矾土
botanical garden　植物园
carnival　*n.* 狂欢节
catalyst　*n.* 催化剂，刺激因素
cater to　迎合，为…服务
citrus　*n.* 柑橘
commoditize　*v.* 商品化
cruise　*n.* 巡游，巡航，乘船游览
curative　*a.* 有疗效的，治病的
distillery　*n.* 酿酒厂
encapsulate　*v.* 概述
entrepreneur　*n.* 企业家
fauna　*n.* 动物群
flora　*n.* 植物群
gear to　使适合于
Gross Domestic Product（GDP）国内生产总值
Gross National Product（GNP）国民生产总值
hurricane　*n.* 飓风
incentive　*n.* 刺激，诱因，激励
infrastructure　*n.* 基础设施
insular　*n.* 海岛的，孤立的，与世隔绝的
itinerary　*n.* 旅行日程
lagoon　*n.* 环礁湖，潟湖；咸水湖
limestone cave　溶洞
lodge　*v.* 暂住，寄存
malaria　*n.* 疟疾，瘴气
microstate　*n.* 微型国家，超小国家
neo-colonization　*n.* 新殖民
oblique　*a.* 倾斜的，不光明正大的
ozone　*n.* 臭氧，新鲜空气
panacea　*n.* 灵丹妙药，万能药
peripheral　*a.* 边缘的，外围的
receipt　*n.* 收入，收据，收到
reggae　*n.* 雷鬼音乐
revenue　*n.* 税收，收益
revitalization　*n.* 复苏，振兴
sanatorium　*n.* 疗养院
scuba diving　轻便潜水

sensuality *n.* 好色，喜爱感官享受
serenity *n.* 宁静，平和
sewage *n.* 排污，下水道
signature dish 招牌菜，拿手菜
snorkeling *n.* 浮潜
sports gear 运动器材
St. Lucia 圣卢西亚岛
stagnant *a.* 停滞的，不景气的
structural handicap 结构性障碍
sustainability *n.* 可持续性
the Balearic Islands 巴利阿里群岛
the Canary Islands 加那利群岛
the Caribbean Islands 加勒比群岛
the Channel Islands 海峡群岛
the Galapagos Islands 加拉帕格斯群岛
the Malvinas Islands 马尔维纳斯群岛
vestige *n.* 遗迹，残余
Virgin Islands 维尔京群岛
Western Samoa 西萨摩亚岛
wildlife reserve 禁猎区，野生动物保护区

Ⅱ. Lead-in & Reading 课文导读

Island Tourism

(1) The concept of island tourism has attracted tourism researchers for ages who tried to examine islands' characteristics, focusing on development within an economic and social context, sustainability and effective management. The physical and climatic characteristics of islands create a particular allure to ever-increasing numbers of tourists. It is this reason that makes islands, and particularly small islands significantly depended on tourism

and as such, tourism has greater economic, socio-cultural and environmental impacts on them than on the mainland.

(2) The concept of the island has received attention by geographers and academics. The geographer Marshall gives the following definition: "and then there are the islands...many are microstates[1] ...vulnerable because of an isolation that produces poverty and instability". Whereas, from the academic point of view, King notes that islands are the "most enticing form of land. Symbol of the eternal contest between land and water...islands suggest mystery and adventure; they inspire and exalt". According to Keane, small islands are attractive for tourists to visit as they create feelings of remoteness and isolation, peace and quiet and sense of timelessness. But many of these factors that constitute the touristic appeal of islands also represent challenges to the longer-term success of tourism related development policies. Islands commonly face a number of structural handicaps arising from their isolated and peripheral location, and their smallness in terms of population and area.

(3) In fact, tourism is often more important for small islands than for mainland destinations, since tourism is invariably a larger and more significant part of the island destination's economy. As such, the ability to understand tourists' needs and to attract a large volume of them is of great importance for the residents, as well as the knowledge of the impacts resulting from tourism on their islands.

(4) Smallness has an important effect on the structure of tourism on islands. In particular, a small island usually implies a less diverse natural resource base. The main resources affected by the competition of tourism seem to be land, water and energy supply, which is considered expensive to produce or import. Moreover, a small population means a limited domestic market. This in turn gives rise to a heavy reliance on foreign trade based on a limited number of products and a narrow range of markets. Climate and geographic isolation works against an economic well-being of the islands. More precisely, the lack of diversity of resources to attract a broad range of international tourists has as a consequence the development of seasonal movements of visitors on the island destinations.

(5) The lack of accessibility as another island characteristic may cause higher transportation cost, decrease in visitors, lack of supplies, high prices

of products and problems in the public services. If an island is not accessible to the outside world, especially in the main tourist trigger countries by air and sea transport, then the development of tourism can only take place on a small scale. That is why many islands are underdeveloped and still have a stagnant socioeconomic structure. However, in the last decade air transport has positively contributed to the opening of new markets.

(6) A common phenomenon in small islands is their undiversified economies. Islands that have one- or two-industry economies (agriculture, fishing, mining, tourism) typically do not have the ability to make substantial investments in tourism marketing or in the creation of a more comprehensive range of products. This has a negative impact on their ability to compete with the thousands of other destinations.

(7) In fact, for island states that have very few resources, virtually the only resources where may be some comparative advantage in favour of are clean beaches, unpolluted seas and warm weather and water, and at least vestiges of distinctive cultures. Indeed, since their ecosystems and natural and scenic beauty are some of their main advantages, islands need to preserve them in order to remain competitive in the tourism industry. This may require that islands place further constraints on travellers, in that they may not be able to do all of the activities that they could do at some other destination. However, protecting the ecosystem makes it potentially more difficult to satisfy the traveller's sense of value.

(8) Tourism has been instrumental in the development of peripheral, remote, and insular regions, like islands, which are commonly characterized by peripherality, isolations, fragility, scarcity of resources, limited labour force and transportation cost, all being competitive disadvantages. Due to the popularity of island destinations, their development is an important and inevitable matter and thus, many islands are using tourism as their major economic growth tool. The development of tourism in many islands comes as a natural consequence; it seems like a common agreement among islands, especially the tropical ones, that they have little economic choice but to accept traditional tourism development and mass tourism as a fact. Many scholars of tourism believe that tourism has become an important vehicle in islands that help them to overcome their size constraints.

(9) Many studies have shown that tourism has become a vital ingredient in many islands' economies. In the island of Antigua for example, tourism receipts account for 58 percent of the Gross National Product (GNP)[2] and in the Canary Islands tourism also accounts for approximately 50 percent of the GDP[3]. Moreover, tourism is also important for many island states because it is a source of foreign exchange.

(10) In addition, perhaps as important as income generation is the employment created by tourism. For instance, in Western Samoa 10 percent of the jobs are tourism related, and in the Malvinas Islands tourism increased the household income by providing part-time jobs. However, most of the jobs for the islanders are usually low on the social and economic scale.

(11) Despite what the critics say, several small islands now use tourism development as a growth strategy to achieve greater economic and development performance, as well as to diversify their economies.

(12) Tourism also improves the standard of living and the infrastructure of the islands. The quality of life and opportunities in the islands may be so poor that many of their inhabitants decide to leave. Tourism, in theoretically providing a mobile and constantly renewable resource, may offer a way of solving some of these problems.

(13) From the social point of view, tourism also contributes to small islands' renewed interest in local arts and crafts, improvements in educational, leisure, communication, medical and other facilities in the host countries, and a broadening in the outlook of the islanders. This is also supported by Dann in his study in St. Lucia, where residents regarded tourism as a catalyst for the awakening or revitalization of local culture.

(14) As the economic potential of tourism has become widely recognized, the preservation of the physical environment has come to be viewed as an investment. Thus, on the positive side, tourism can be credited with extending environmental appreciation and providing an incentive for conservation policies; many island governments are placing greater emphasis now on "clean up" campaigns.

(15) Tourism is not a panacea for all the problems of the islands and apart from the positive effects there are also other adverse impacts of tourism development. The general problem essentially is one of too many tourists who

provide jobs for too many residents on a group of islands too small to lodge everyone.

(16) Island tourism development is widely considered to be typified by, amongst other factors, small geographical size, distance and isolation from metropolitan centres, a limited economic base, and a lack of resources. As a consequence island economies often become dependent upon a dominant tourism sector which is controlled by overseas tour operators, airlines and hotel chains. This situation is often referred to in the literature as neo-colonization[4]. For instance, out of sixty world destinations where tourism receipts are highest relative to total national revenue, the top fifteen are islands, and such evidence highlights the level of dependency on tourism experienced by small island states.

(17) Many islands are developed by the tourism sector only for seasonal tourism. Seasonal fluctuations result in seasonal employment, under-utilisation of the facilities, overcrowding and stresses upon the transport system and the public services.

(18) Tourism development often contrasts sharply with the protection of uniqueness, as it implies modernisation, change in culture, urbanization and exploitation of resources. The phenomenon whereby locals begin to mimic the culture of their visitors, thereby diminishing the importance and permanence of their own culture and heritage, is well recognized as a cultural impact concern. Furthermore, young people see tourists as role models, arts and crafts become commoditized, songs and dances are geared to commercial performance, changing to cater for tourist taste and meet the constraints of the tourist itinerary. Many times, locals overwork during the summer season, altering or even neglecting their social, family, religious and cultural obligations. One of the most popular subjects in island tourism is the impact of tourism on the islands environment. Tourism very often destroys the serenity and beauty of islands through congestion, traffic, and unpleasant development. Small islands have small social and environmental carrying capacities, and so the adverse impacts of tourism tend to be more severe in these places than in the large mainland countries. The fragility of their ecosystems arises as a result of a low level of resistance to outside influences.

(19) Tourist development on islands leads to destruction of native

forests, flora and fauna during airport, port and hotel construction and damage to beaches; Moreover, once established, hotels and other tourist sites may pollute lagoons and water through the discharge of waste and sewage, and the activities of tourists themselves may prompt further damage. More precisely, coastal pollution, water shortages, sewage treatment, waste disposal, traffic congestion, noise pollution, overbuilding, and aesthetic degradation are some of the impacts that different studies found in the Greek islands, the Balearic Islands, the Caribbean, the Galapagos and the Channel Islands.

(20) The current study encapsulated the main characteristics of island tourism and presented the impacts that tourism development has on the islands worldwide. Tourism on islands is not a simplified subject. Islands are frequently characterized by smallness, insularity, undiversified economies and fragile ecosystems, and thus they face particular disadvantages for tourism in this competitive environment. Moreover, as the decision to visit an island is influenced by conditions outside of the control of the island itself, islands are vulnerable to often minor fluctuations in world market conditions and their economies have tended to be heavily distorted and unstable.

Notes:

[1] microstate　微型国家或超小国家，又名 ministate，是指人口很少或者面积很小的主权国家。其中最小的是梵蒂冈。根据 2013 年的统计，梵蒂冈只有公民 842 人，面积为 0.44 平方千米。

[2] GNP　国民生产总值（Gross National Product），是指一个国家（或地区）所有国民在一定时期内新生产的产品和服务价值的总和。GNP 是按国民原则核算的，只要是本国（或地区）居民，无论是否在本国境内（或地区内）居住，其生产和经营活动新创造的增加值都应该计算在内。例如，中国的居民通过劳务输出在境外所获得的收入就应该计算在中国的 GNP 中。

[3] GDP　国内生产总值（Gross Domestic Product），是指一个国家（国界范围内）所有常驻单位在一定时期内生产的所有最终产品和劳务的市场价值。GDP 是国民经济核算的核心指标，也是衡量一个国家或地区总体经济状况重要指标。

[4] neo-colonization　新殖民，是新殖民主义（neo-colonialism）的表现形式。新殖民主义是指第二次世界大战后殖民主义体系崩溃，在新形势下殖民主义者为维护旧殖民利益并获取新的利益提出的思想理论体系以及采取的新的殖民手段和政策。

Ⅲ. Teaching Focus (7): Longish and Complicated Sentences 翻译技巧 (7): 长句与复杂句

英语和汉语属于不同的语系，在句子结构上有很大的不同。汉语句子多为并列结构，而英语句子多为主从结构，有很多长句与复杂句，科技语篇尤其如此，往往是技术性越强，越正式，句子就越长。叠床架屋，错综复杂，有时一个句子就是长长一段话。不仅并列句和复合句如此，有时简单句也会很长。英语语言的特点也使这种长而复杂的句子成为可能，因为英语修饰语的位置很灵活，很容易扩展。此外，英语中有各种灵活的连接手段，可以把各种并列、修饰或附加成分组合起来。因此，英语句子呈现句首封闭、句尾开放（right-branching）的特征，修饰、并列和附加成分都可以后置，句子可以不断向句尾延伸。

对于英语学习者来说，这样的长句子让人头晕眼花，望而生畏。实际上，对这样的句子，只要保持冷静，宏观把握整个句子，确定句子类型，是简单句、复杂句、复合句还是包括很多从句的复合复杂句，然后依据意群对长句进行分割，确定句子的主干（通常为主谓或主谓宾结构）和分枝。这一过程如同抽丝剥茧，或者说是像剥洋葱一样，只要弄清句子成分之间的逻辑关系，找到语义重心所在，正确理解并不难。

对于译者来说，关键的是第二环节，即用符合目标语习惯的方式将原文所表达的意义表述出来。相对英语的繁复，汉语结构简单，有很多省略，以中短句居多。书面语中虽然也有结构复杂的长句，但常用标点或虚词分开。这是因为汉语中的修饰成分一般前置，一个词语所能承受的修饰成分有限。换句话说，汉语句子呈现句首开放、句尾收缩（left-branching）的特征。其扩展的长度和程度受到种种限制，不能像英语那样层层修饰，向后不断扩展。因此，在进行英译汉时，要特别注意英语和汉语之间的差异，将英语的长句进行分解，按照汉语的特点和表达习惯，化整为零，正确地译出原文的意思。在此过程中，重点是要跳出英语的句法框架，不能拘泥于原文的形式。就技巧而言，这里介绍 4 种常用的方法并结合课文导读中的句子分别举例说明。

1. 顺序法 在英语句子的句法结构和逻辑顺序与汉语相同或相近时，可以用和原句相同的顺序来组织汉语句子，这和口译中的“顺句驱动”比较接近。由于英语长句是由各种不同的从句和短语构成的，通过词性的变化和连词的运用来表达这些成分之间的逻辑关系，在分解成汉语短句时，应特别注意汉语连词的使用。

【例 1】From the social point of view，tourism also contributes to small islands' renewed interest in local arts and crafts，improvements in educational，leisure，communication，medical and other facilities in the host countries，and a broadening in the outlook of the islanders.

【译文】从社会的角度来看，旅游业也让小岛屿重新对当地的艺术和工艺产生兴趣，改进东道主国的教育、休闲、通讯、医疗和其他设施，开阔岛上居民的眼界。

【分析】本例主干是一个主谓宾结构，翻译的难点在于谓语动词"contribute"和三个并列的宾语成分。英语初学者常犯的一个错误就是看到"contribute"或"contribution"马上就想到"贡献"，看到"让""使"就马上想到"make"或"let"。在翻译宾语成分时，一定要注意将英语中名词转换成汉语中的动词。

【例 2】For instance，out of sixty world destinations where tourism receipts are highest relative to total national revenue，the top fifteen are islands，and such evidence highlights the level of dependency on tourism experienced by small island states.

【译文】例如，在世界上旅游业收入占国家总收入比例最高的 60 个旅游目的地中，前 15 名都是岛屿。由此可见小岛屿国家对旅游业的依赖程度。

【分析】本例主干是"and"连接的两个并列句，前者是主系表结构，后者是主谓宾结构。前者的难点是定语从句和名词"receipt"的翻译，后者的难点在于谓语动词"highlight"的表达。

【例 3】Furthermore，young people see tourists as role models，arts and crafts become commoditized，songs and dances are geared to commercial performance，changing to cater for tourist taste and meet the constraints of the tourist itinerary.

【译文】此外，年轻人会模仿游客，艺术和工艺被商品化，歌舞被用于商业性的演出，为了迎合游客的品味、满足游客的行程需求而做出改变。

【分析】本例的主干是三个并列成分，难度并不大，重点要注意动词短语"gear to"和"cater for"以及"meet the constraints"的译法。

2. 逆序法 逆序法又名反向翻译法，如果英语长句的结构顺序和与其相对应的汉语句子差异很大，甚至恰恰相反，就要使用这种方法。

【例 4】More precisely，coastal pollution，water shortages，sewage treatment，waste disposal，traffic congestion，noise pollution，overbuilding，and aesthetic degradation are some of the impacts that different studies found in the Greek islands，the Balearic Islands，the Caribbean，the Galapagos and the Channel Islands

【译文】精确说来，在希腊群岛、巴利阿里群岛、加勒比群岛、加拉帕格斯群岛和海峡群岛所进行的不同研究发现了下列一些影响：沿海污染、水短缺、污水和垃圾处理、交通拥堵、噪音污染、过度建筑和审美退化。

【分析】除了句首的副词成分之外，本句是一个典型的从后往前翻的例子。本句的主干是一个主系表结构，主语由八个并列的名词短语组成，表语后面有一个很长的定语从句，并且从句中又有五个并列的名词短语。如果按照英语顺序逐词翻译出来，会冗长拖沓，佶屈聱牙，不堪卒读。

3. 拆分法 拆分法是把一个长而复杂的句子拆译成若干个较短、较简单的句子，通常用于英译汉。

【例 5】Islands are frequently characterized by smallness，insularity，undiversified economies and fragile ecosystems，and thus they face particular disadvantages for tourism in this competitive environment.

【译文】岛屿经常有如下特征：面积小，与世隔绝，经济单一，生态系统脆弱。因此，在这个竞争激烈的环境中，它们在发展旅游业方面尤其面临劣势。

【分析】本例的重点在于短语“be characterized by”和其后平行结构的翻译，对于前者，翻译策略是动词名词化，而在翻译英语的平行结构时，并列的名词成分被翻译成一并列的短语或短句。此外还要注意确保语法形式的对应，即名词对名词，动词对动词，形容词对形容词，副词对副词，分词短语对分词短语，分句对分句，只有这样，才能保留原文的节奏和气势，读起来抑扬顿挫，铿锵有力。

4. 插入法 如果重新组织原句中分散的修饰成分有困难，可以借助括号或破折号将其插入汉语句子中。当然，如果源语句子中有括号，可以直接将里面的内容翻译到目标语的括号内。有的英语句子中虽然没有括号，但是在翻译时有些附加成分难以通过其他方式处理，在目标语中把这些信息加上括号是不错的选择。

【例 6】It is this reason that makes islands，and particularly small islands significantly depended on tourism and as such，tourism has greater economic，socio-cultural and environmental impacts on them than on the mainland.

【译文】正是这一原因让岛屿（尤其是小岛屿）对旅游业十分依赖。因此，旅游业对其经济、社会、文化和环境的影响要大于大陆。

【分析】本例是由连词“and”连接的两个并列句，前半句是常见的强调结构，后半句为主谓结构。在前半句中，“and particularly small islands”属于插入成分，表示补充说明，如果直接翻译过来，不符合汉语的句法习惯，因此最好加一个括号。后半句中多个形容词并列，难度并不大，问题在于后面比较结构的表达。

【例7】Islands that have one- or two-industry economies（agriculture，fishing，mining，tourism）typically do not have the ability to make substantial investments in tourism marketing or in the creation of a more comprehensive range of products.

【译文】只有一两种产业（农业、渔业、矿业或旅游业）的岛屿无法大量投资于旅游业的推广，也不能提供更加全面多样的产品。

【分析】原文括号里的内容是对前面内容的具体说明，因此翻译过来之后直接保留括号即可。此句的主干是一个主谓宾结构，即islands do not have the ability。其他的定语从句、副词、不定式和介词短语都是修饰和限定成分。注意这里“marketing”如果翻译成“营销”，有点不符合汉语表达习惯，可以译为“推广”。同理，“creation”被灵活地译为“提供”，而不是“创造”。

【例8】The development of tourism in many islands comes as a natural consequence；it seems like a common agreement among islands，especially the tropical ones，that they have little economic choice but to accept traditional tourism development and mass tourism as a fact.

【译文】在很多岛屿上，旅游业的发展是自然发生的结果。岛屿之间——尤其是热带的岛屿之间——似乎达成了某种共识，即在经济上，它们别无选择，只有接受传统的旅游业发展和大众旅游业。

【分析】本例为分号连接的两个分句，前面是一个主谓结构，后面是形式主语“it”所引导的主语从句，真正的主语是“that”所引导的从句。“especially the tropical ones”属于插入成分，表示补充说明，翻译时可以放入括号之内，也可以前后加破折号。

总之，译无定法，贵在变通。无论是顺序法或逆序法，还是拆分法或插入法，这些都是为了翻译研究和教学的方便而发明的描述性表达，因此要灵活把握，综合运用，而不能被其约束。正如拳术，一招一式，目的在于练习，练得熟练了，运用时就会驾轻就熟。否则，在实战中，如果总是想着该用哪个招式，就只有被动挨打。作为译者，对于源语，要做到“得意忘言”，甚至要“得意忘形”，务必要遵守目标语的表达习惯，只有这样，才能避免所谓的“翻译腔”。

Questions for Discussion:

1. What are the differences between Chinese and English in terms of syntactic structure?

2. How to deal with the syntactic differences in translation?

Ⅳ. Exercise 1: Sentence in Focus　练习1：单句翻译

1. Translate the following sentences into Chinese

(1) The physical and climatic characteristics of islands create a particular allure to ever-increasing numbers of tourists.

(2) The lack of diversity of resources to attract a broad range of international tourists has as a consequence the development of seasonal movements of visitors on the island destinations.

(3) But many of these factors that constitute the touristic appeal of islands also represent challenges to the longer-term success of tourism related development policies.

(4) The ability to understand tourists' needs and to attract a large volume of them is of great importance for the residents, as well as the knowledge of the impacts resulting from tourism on their islands.

(5) Climate and geographic isolation works against an economic well-being of the islands.

(6) The lack of accessibility as another island characteristic may cause higher transportation cost, decrease in visitors, lack of supplies, high prices of products and problems in the public services.

(7) Indeed, since their ecosystems and natural and scenic beauty are some of their main advantages, islands need to preserve them in order to remain competitive in the tourism industry.

(8) Islands are commonly characterized by peripherality, isolations, fragility, scarcity of resources, limited labour force and transportation cost, all being competitive disadvantages.

(9) Many scholars of tourism believe that tourism has become an important vehicle in islands that help them to overcome their size constraints.

(10) Tourism, in theoretically providing a mobile and constantly

renewable resource, may offer a way of solving some of these problems.

(11) Tourism is not a panacea for all the problems of the islands and apart from the positive effects there are also other adverse impacts of tourism development.

(12) Thus, on the positive side, tourism can be credited with extending environmental appreciation and providing an incentive for conservation policies.

(13) Seasonal fluctuations result in seasonal employment, under-utilisation of the facilities, overcrowding and stresses upon the transport system and the public services.

(14) The phenomenon whereby locals begin to mimic the culture of their visitors, thereby diminishing the importance and permanence of their own culture and heritage, is well recognized as a cultural impact concern.

(15) As the decision to visit an island is influenced by conditions outside of the control of the island itself, islands are vulnerable to often minor fluctuations in world market conditions.

2. Translate the following sentences into English

(1) 工业化捕捞的到来让很多海洋物种面临灭顶之灾。

(2) 对于度蜜月者来说，南太平洋的很多岛屿有无法抵抗的吸引力。

(3) 不负责任的捕捞方式是危害渔业可持续性的主要力量。

(4) 中国政府正着力减少汽车尾气排放，汽车制造商为了迎合环境友好汽车这种需要而积极努力。

(5) 这个国家公园的重要性来自它丰富的植物和动物物种。

(6)《华尔街日报》的一篇社论概述了很多保守派人士的观点。

(7) 这一激励体制根据个人表现和单个目标的实现对个体给予奖励。

(8) 这里的很多处温泉有治疗效果，其中一个每分钟可以排出大约 3 000 加仑的泉水。

(9) 水圈实际上是连接在一起的，而陆地却互相孤立，其中最大的是亚非欧的旧世界，其次是美洲，南极洲可能排第三位，排在第四位的是澳洲。

(10) 公司越来越倾向于把像培训这样的非核心业务外包出去。

(11) 在白鲸自然保护区，人们可以与这些奇妙的生物游泳甚至拥抱它们。

(12) 旅游业收入是该国无形收益当中最大的一项。

(13) 要想振兴旅游业，首先要重视生态环境的保护。

(14) 经济停滞不前，加上青少年人数激增，可能是导致美国犯罪率上升的原因之一。

（15）自然有一种神奇的力量，能够让即使是最浮躁的心平静下来，给人祥和与宁静。

Ⅴ. Exercise 2：Passages in Focus　练习 2：短文翻译

1. Translate the following passage into Chinese

Caribbean islands now depend on tourism for their economy，it being referred to as "the engine of their growth". Tourism is a huge contributor to the economies of all Caribbean countries and the biggest contributor to many of them such as Antigua and Barbuda，Bahamas and the Virgin Islands. It provides a steady revenue stream，with temporary blips due to hurricanes or recessions in the Western world and supports local farming，fishing，and retail industries. Barbados，for instance，has moved from a primarily agricultural economy to a service-based economy that supports tourism. By 2006，tourism brought in ten times more to the Barbados economy than sugar cane production，$167 million versus $14.5 million. In terms of employment，11.3% of the region's jobs depend on tourism either directly or indirectly. It is often described as "the most tourism-dependent region in the world".

Tourist attractions of the region are those generally associated with a maritime tropical climate：Scuba diving and snorkeling on coral reefs，cruises，sailing，and game fishing at sea. On land：golf，botanical gardens，parks，limestone caves，wildlife reserves，hiking，cycling and horseback riding. Cultural attractions include Carnival，steel bands，reggae and cricket. Due to the dispersal of the islands，helicopter or aeroplane tours are popular. Specific to this tropical region are tours of historic colonial plantation houses，sugar mills and rum distilleries. Caribbean cuisine is a fusion of cooking styles，goat stew being the signature dish of several islands.

A large number of the visitors are honeymooners or people who come to the islands for a destination wedding. The early seaside resorts were developed primarily for curative benefits of bathing in the sea and breathing the warm，ozone-laden air. Barbados was referred to as the "sanatorium of the West Indies" in guidebooks because of its fresh water，sea air，and absence of malaria.

In the 1920s, tourists visited the Caribbean for pleasurable, sun-bathing vacations. Sun exposure was considered healthy at that time and tans were a symbol of sensuality among the wealthy. Before World War Ⅱ, more than 100,000 tourists visited the region a year.

Tourism became an economically important industry as Caribbean bananas, sugar, and bauxite were no longer competitively priced with the advent of free-trade policies. Encouraged by the United Nations and World Bank, many governments in the Caribbean encouraged tourism beginning in the 1950s to boost the developing countries economies. The Caribbean Tourist Association was founded in 1951. Tax incentives encouraged foreign development of hotels and infrastructure, cultivated by newly formed tourism ministries.

2. Translate the following passage into English

作为小岛屿国家联盟（Alliance of Small Islands States）的成员，马尔代夫群岛是最容易受气候变化影响的国家之一。这是由于马尔代夫群岛和该联盟其他国家所共有的地理和经济特征而造成的。它们都是被大海所包围的小块陆地，常常位于易发自然灾害的地区。此外，它们的基础设施不够发达，资源有限，在经济上依附于其他国家。

就马尔代夫群岛而言，其经济和人民的福利非常依赖于良好的环境条件，任何气候条件的变化都可能是危险的。旅游业直接或间接地占财政收入的很大一部分。此外，20%的人口主要通过渔业获得收入，并且岛上所有的基础设施都仅高于海平面 1.5 米。

因此，环境模式的任何变化都可以被看作是对马尔代夫人的威胁，例如 1987 年的洪水，给马累国际机场造成了 450 万美元的经济损失，而受 2004 年海啸的影响，旅游业损失高达 2.3 亿美元。除了结构上的损坏之外，还有的威胁到马尔代夫人的生命安全。降水模式的变化和洪水造成疾病的多发，如登革热之类的。由于气温和降水的变化，马尔代夫人每年都会经历长达十多天的缺水。

可见，虽然马尔代夫群岛和其他的小岛国一样，是温室气体排放最少的国家之一，占全球排放总量不到 1%，却是最容易受其影响的国家之一。现在，政府不得不将资源投入应对气候变化带来的恶果，而不是用于脱贫和改进民生。

对于海平面上升，学者之间的探讨很激烈。由于分析这一情况的方法不同，对于气候变化和海平面上升之间的关系并没有达成共识。但是，由于本世纪平均海平面的上升，马尔代夫群岛的确面临着更大的被淹没的危险，加剧了海平面的短暂变化和作为侵蚀主要原因的风浪和涌浪所带来的现有威胁。

Unit Eight Marine Laws and Regulations

第八单元　海洋法律法规

Ⅰ. Warm-up for Theme-related Words/Expressions　词汇预习

adjacent　*a.* 邻近的，毗邻的
archipelagic　*a.* 群岛的
armada　*n.* 舰队
Bab el Mandeb　曼德海峡
beneficiary　*n.* 受益者
contiguous　*a.* 相连的，接触的
cottage industry　家庭手工业
depletion　*n.* 耗竭，消耗，用尽
designation　*n.* 指定
detour　*n.* 绕道，迂回
detrimental　*a.* 有害的，不利的
dhow　*n.* 单桅三角帆船
dominion　*n.* 主权，统治权
endowment　*n.* 赠与，天赋
escort　*v.* 护送，护航
expeditious　*a.* 迅速的，敏捷的
eye　*v.* 注意，看
flotilla　*n.* 小型船队
husband　*v.* 节约地使用或管理
hydrocarbon reserve　油气储量
infringement　*n.* 侵犯，违反
intruder　*v.* 侵入者
jurisdiction　*n.* 管辖权，司法权
league　*n.* 里格（长度单位）
Lima　*n.* 利马
littoral　*a.* 沿海的，滨海的
lucrative　*a.* 赚钱的，有利可图的
Mediterranean　*n.* 地中海
Montevideo　*n.* 蒙得维的亚
mutatis mutandis　加上必要的变更
optimum　*a.* 最佳的
overflight　*n.* 飞越上空
papyrus　*n.* 纸莎草
patrol　*v.* 巡逻，巡航
ply　*v.* 定期来往于
prejudicial　*a.* 有害于，不利于
projectile　*n.* 投射物
proximity　*n.* 接近，邻近
regime　*n.* 管理体系，制度，政权
riparian　*a.* 沿岸的
smuggler　*n.* 走私者，走私船
sovereign　*n.* 主权国
Strait of Gibraltar　直布罗陀海峡
Strait of Hormuz　霍尔木兹海峡
Strait of Malacca　马六甲海峡
transient　*a.* 路过的，短暂的
transit　*n.* 经过，运输
unilateral　*a.* 单边的

unimpeded *a.* 不受阻碍的　　　　vie *v.* 竞争

Ⅱ. Lead-in & Reading 课文导读

The Law of the Sea

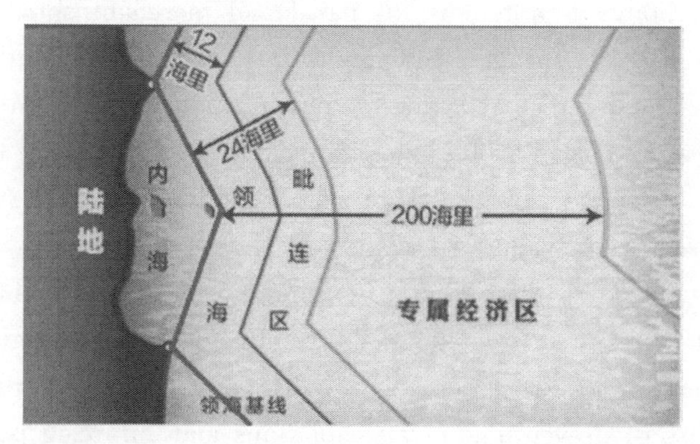

(1) The dispute over who controls the oceans probably dates back to the days when the Egyptians first plied the Mediterranean in papyrus rafts. Over the years and centuries, countries large and small, possessing vast ocean-going fleets or small fishing flotillas, husbanding rich fishing grounds close to shore or eyeing distant harvests, have all vied for the right to call long stretches of oceans and seas of their own.

(2) Before the Convention on the Law of the Sea could address the exploitation of the riches underneath the high seas, navigation rights, economic jurisdiction, or any other pressing matter, it had to face one major and primary issue—the setting of limits. Everything else would depend on clearly defining the line separating national and international waters. Though the right of a coastal State to complete control over a belt of water along its shoreline—the territorial sea—had long been recognized in international law, up until the Third United Nations Conference on the Law of the Sea, States could not see eye to eye on how narrow or wide this belt should be.

(3) Traditionally, smaller States and those not possessing large, ocean-going navies or merchant fleets favoured a wide territorial sea in order to protect their coastal waters from infringements by those States that did. Naval and maritime Powers, on the other hand, sought to limit the territorial sea as

much as possible, in order to protect their fleets' freedom of movement.

(4) As the work of the Conference progressed, the move towards a 12-mile territorial sea gained wider and eventually universal acceptance. Within this limit, States are in principle free to enforce any law, regulate any use and exploit any resource.

(5) The Convention retains for naval and merchant ships the right of "innocent passage"[1] through the territorial seas of a coastal State. This means, for example, that a Japanese ship, picking up oil from Gulf States, would not have to make a 3,000-mile detour in order to avoid the territorial sea of Indonesia, provided passage is not detrimental to Indonesia and does not threaten its security or violate its laws.

(6) In addition to their right to enforce any law within their territorial seas, coastal States are also empowered to implement certain rights in an area beyond the territorial sea, extending for 24 nautical miles from their shores, for the purpose of preventing certain violations and enforcing police powers. This area, known as the "contiguous zone"[2], may be used by a coast guard or its naval equivalent to pursue and, if necessary, arrest and detain suspected drug smugglers, illegal immigrants and customs or tax evaders violating the laws of the coastal State within its territory or the territorial sea.

(7) The Convention also contains a new feature in international law, which is the regime for archipelagic States (States such as the Philippines and Indonesia, which are made up of a group of closely spaced islands). For those States, the territorial sea is a 12-mile zone extending from a line drawn joining the outermost points of the outermost islands of the group that are in close proximity to each other. The waters between the islands are declared archipelagic waters, where ships of all States enjoy the right of innocent passage. In those waters, States may establish sea lanes and air routes where all ships and aircraft enjoy the right of expeditious and unobstructed passage.

Navigation

(8) Perhaps no other issue was considered as vital or presented the negotiators of the Convention on the Law of the Sea with as much difficulty as that of navigational rights.

(9) Countries have generally claimed some part of the seas beyond their

shores as part of their territory, as a zone of protection to be patrolled against smugglers, warships and other intruders. At its origin, the basis of the claim of coastal States to a belt of the sea was the principle of protection; during the seventeenth and eighteenth centuries another principle gradually evolved: that the extent of this belt should be measured by the power of the littoral sovereign to control the area.

(10) In the eighteenth century, the so-called "cannon-shot" rule gained wide acceptance in Europe. Coastal States were to exercise dominion over their territorial seas as far as projectiles could be fired from a cannon based on the shore. According to some scholars, in the eighteenth century the range of land-based cannons was approximately one marine league, or three nautical miles. It is believed that on the basis of this formula developed the traditional three-mile territorial sea limit.

(11) By the late 1960s, a trend to a 12-mile territorial sea had gradually emerged throughout the world, with a great majority of nations claiming sovereignty out to that seaward limit. However, the major maritime and naval Powers clung to a three-mile limit on territorial seas, primarily because a 12-mile limit would effectively close off and place under national sovereignty more than 100 straits used for international navigation.

(12) A 12-mile territorial sea would place under national jurisdiction of riparian States strategic passages such as the Strait of Gibraltar (8 miles wide and the only open access to the Mediterranean), the Strait of Malacca (20 miles wide and the main sea route between the Pacific and Indian Oceans), the Strait of Hormuz (21 miles wide and the only passage to the oil-producing areas of Gulf States) and Bab el Mandeb (14 miles wide, connecting the Indian Ocean with the Red Sea).

(13) At the Third United Nations Conference on the Law of the Sea, the issue of passage through straits placed the major naval Powers on one side and coastal States controlling narrow straits on the other. The United States and the Soviet Union insisted on free passage through straits, in effect giving straits the same legal status as the international waters of the high seas. The coastal states, concerned that passage of foreign warships so close to their shores might pose a threat to their national security and possibly involve them in conflicts among outside Powers, rejected this demand.

(14) Instead, coastal States insisted on the designation of straits as territorial seas and were willing to grant to foreign warships only the right of "innocent passage", a term that was generally recognized to mean passage "not prejudicial to the peace, good order or security of the coastal State". The major naval Powers rejected this concept, since, under international law, a submarine exercising its right of innocent passage, for example, would have to surface and show its flag—an unacceptable security risk in the eyes of naval Powers. Also, innocent passage does not guarantee the aircraft of foreign States the right of overflight over waters where only such passage is guaranteed.

(15) The compromise that emerged in the Convention is a new concept that combines the legally accepted provisions of innocent passage through territorial waters and freedom of navigation on the high seas. The new concept, "transit passage"[3], required concessions from both sides.

(16) The regime of transit passage retains the international status of the straits and gives the naval Powers the right to unimpeded navigation and overflight that they had insisted on. Ships and vessels in transit passage, however, must observe international regulations on navigational safety, civilian air-traffic control and prohibition of vessel-source pollution and the conditions that ships and aircraft proceed without delay and without stopping except in distress situations and that they refrain from any threat or use of force against the coastal State. In all matters other than such transient navigation, straits are to be considered part of the territorial sea of the coastal State.

Exclusive Economic Zone

(17) The exclusive economic zone (EEZ) is one of the most revolutionary features of the Convention, and one which already has had a profound impact on the management and conservation of the resources of the oceans. Simply put, it recognizes the right of coastal States to jurisdiction over the resources of some 38 million square nautical miles of ocean space. To the coastal State falls the right to exploit, develop, manage and conserve all resources to be found in the waters, on the ocean floor and in the subsoil of an area extending 200 miles from its shore.

(18) The EEZs are a generous endowment indeed. About 87 percent of all known and estimated hydrocarbon reserves under the sea fall under some national jurisdiction as a result. So too will almost all known and potential offshore mineral resources, excluding the mineral resources of the deep ocean floor beyond national limits.

(19) The most lucrative fishing grounds too are predominantly the coastal waters. This is because the richest phytoplankton pastures lie within 200 miles of the continental masses. Phytoplankton, the basic food of fish, is brought up from the deep by currents and ocean streams at their strongest near land, and by the upwelling of cold waters where there are strong offshore winds.

(20) The desire of coastal States to control the fish harvest in adjacent waters was a major driving force behind the creation of the EEZs. Fishing, the prototypical cottage industry before the Second World War, had grown tremendously by the 1950s and 1960s. Fifteen million tons in 1938, the world fish catch stood at 86 million tons in 1989. No longer the domain of a lone fisherman plying the sea in a wooden dhow, fishing, to be competitive in world markets, now requires armadas of factory-fishing vessels, able to stay months at sea far from their native shores, and carrying sophisticated equipment for tracking their prey.

(21) The special interest of coastal States in the conservation and management of fisheries in adjacent waters was first recognized in the 1958 Convention on Fishing and Conservation of the Living Resources of the High Seas. That Convention allowed coastal States to take "unilateral measures" of conservation on what was then the high seas adjacent to their territorial waters. It required that if six months of prior negotiations with foreign fishing nations had failed to find a formula for sharing, the coastal State could impose terms. But still the rules were disorderly, procedures undefined, and rights and obligations a web of confusion. On the whole, these rules were never implemented.

(22) The claim for 200-mile offshore sovereignty made by Peru, Chile and Ecuador in the late 1940s and early 1950s was sparked by their desire to protect from foreign fishermen the rich waters of the Humboldt Current (more or less coinciding with the 200-mile offshore belt). This limit was incorporated in the Santiago Declaration of 1952 and reaffirmed by other Latin

American States joining the three in the Montevideo and Lima Declarations of 1970. The idea of sovereignty over coastal-area resources continued to gain ground.

(23) As long-utilized fishing grounds began to show signs of depletion, as long-distance ships came to fish waters local fishermen claimed by tradition, as competition increased, so too did conflict. Between 1974 and 1979 alone there were some 20 disputes over cod, anchovies or tuna and other species between, for example, the United Kingdom and Iceland, Morocco and Spain, and the United States and Peru.

(24) Today, the benefits brought by the EEZs are more clearly evident. Already 86 coastal States have economic jurisdiction up to the 200-mile limit. As a result, almost 99 percent of the world's fisheries now fall under some nation's jurisdiction. Also, a large percentage of world oil and gas production is offshore. Many other marine resources also fall within coastal-State control. This provides a long-needed opportunity for rational, well-managed exploitation under an assured authority.

(25) Figures on known offshore oil reserves now range from 240 to 300 billion tons. Production from these reserves amounted to a little more than 25 percent of total world production in 1996. Experts estimate that of the 150 countries with offshore jurisdiction, over 100, many of them developing countries, have medium to excellent prospects of finding and developing new oil and natural gas fields.

(26) It is evident that it is archipelagic States and large nations endowed with long coastlines that naturally acquire the greatest areas under the EEZ regime. Among the major beneficiaries of the EEZ regime are the United States, France, Indonesia, New Zealand, Australia and the Russian Federation.

(27) But with exclusive rights come responsibilities and obligations. For example, the Convention encourages optimum use of fish stocks without risking depletion through overfishing. Each coastal State is to determine the total allowable catch for each fish species within its economic zone and is also to estimate its harvest capacity and what it can and cannot itself catch. Coastal States are obliged to give access to others, particularly neighbouring States and land-locked countries, to the surplus of the allowable catch. Such access

must be done in accordance with the conservation measures established in the laws and regulations of the coastal State.

(28) Coastal States have certain other obligations, including the adoption of measures to prevent and limit pollution and to facilitate marine scientific research in their EEZs.

Notes:

[1] innocent passage 无害通过，是指外国（或地区）船舶在不损害沿海国（或地区）的和平、安全和良好秩序条件下，继续不停地迅速通过沿海国领海的航行。

[2] contiguous zone 毗连区，是指沿海国领海以外毗邻领海，由沿海国对其海关、财政、卫生和移民等事项行使管辖权的一定宽度的海洋区域。毗连区的宽度从领海基线量起不超过24海里。

[3] transit passage 过境通行，是指外国船舶和飞机自由地（不仅是“无害地”）通过用于国际航行的海峡（即使该海峡完全处在沿岸国的领海范围以内），由公海的一个海域或一个专属经济区到公海的另一个海域或另一个专属经济区的航行。

Ⅲ. Teaching Focus (8): Conversion of Voice 翻译技巧(8)：语态转换

作为动词的一种句法形式，语态被用来表达主语与宾语之间的行为关系。当动作执行者作主语时，句子结构就是主动语态；当动作的承受者作主语时，句子结构就是被动语态。如果将汉英两种语言进行比较，会发现英语中的被动语态的句子远远多于汉语，汉语中习惯多用主动语态，大量使用无主句，并且多用主动结构表达被动意义。英语中的被动语态是以动词的形态变化作为标记的，而汉语的动词则没有曲折变化体系，只能利用词汇手段来表达被动。因此，这里所说的语态转换法主要是英语里被动语态在被翻译成汉语时的处理方法。

科技英语的特点之一就是被动语态的广泛使用，因为和主动语态相比，被动语态更能突出所要论证的对象。此外，被动语态的主观色彩比主动语态更少，而这正是科技文体所需要的特征，涉海科技英语也不例外。因此，海洋英语文献中会有很多被动语态的句子，给译者的理解和翻译带来一些困难，如英语中的被动语态常被翻译成汉语的主动语态，或者是无主句。一般说来，对英语被动语态的汉译有以下几种方法：

1. 顺译法 顺译法即保留原文的主语，按照顺句驱动的原则，使译文主要成分的顺序基本上和原文一致。

(1) 顺译成被动句。

【例 1】The waters between the islands are declared archipelagic waters, where ships of all States enjoy the right of innocent passage.

【译文】岛屿之间的水域被宣称为群岛水域，在这里，各国的船只都享有无害通过权。

【分析】原文被翻译成了汉语里的“被”字句，即在动词前面加上一个“被”字来表达被动意义。同样可以用来表达被动意义的词语还有“给、让、由、受、遭到、受到、予以、加以、得以、为…所”等，如下面几个例子：

【例 2】On the whole, these rules were never implemented.

【译文】总体上看，这些规定从来没有得以执行。

【例 3】Though the right of a coastal State to complete control over a belt of water along its shoreline had long been recognized in international law, up until the Third United Nations Conference on the Law of the Sea, States could not see eye to eye on how narrow or wide this belt should be.

【译文】虽然沿海国家完全支配沿海岸线水域带的权利早已为国际法所承认，在第三次联合国海洋法会议召开之前，各国并未能就这条水域的宽度达成一致。

【例 4】The extent of this belt should be measured by the power of the littoral sovereign to control the area.

【译文】这条水域的范围应该由沿海主权国家控制这一区域的能力来衡量。

（2）顺译成主动句。

【例 5】Coastal States are obliged to give access to others, particularly neighbouring States and land-locked countries, to the surplus of the allowable catch. Such access must be done in accordance with the conservation measures established in the laws and regulations of the coastal State.

【译文】沿海国必须把可捕获量中过剩的那部分让给其他国家捕捞，尤其是邻国和内陆国，而在捕捞的过程中，它们必须遵守由沿海国的法律和规定所确立的养护措施。

【分析】原文由两个被动句组成，而翻译成汉语之后，却看不到被动的痕迹，这是因为英语中有些被动语态的句子，如果生硬翻译成被动句，就会不符合汉语的表达习惯。英语里的“be obliged to”虽然有被迫的意思，但是常被作为固定用法，表示有义务做某事，不得不做某事，这里被翻译成了“必须”。在翻译原文第二句时，主语和谓语动词都发生了变化。当然，access 一词增加了这句话的翻译难度。

【例 6】The claim for 200-mile offshore sovereignty made by Peru, Chile and Ecuador in the late 1940s and early 1950s was sparked by their desire to protect from foreign fishermen the rich waters of the Humboldt Current.

【译文】20 世纪 40 年代末和 50 年代初，秘鲁、智利和厄瓜多尔提出 200 海里海上主权的主张，就是希望能够保护洪堡洋流资源丰富的海域使其不受外国渔民的入侵。

【分析】本例主语后面的限定成分太长，被翻译成相对独立的分句，相应地，原文的谓语部分也要做出调整，表示被动意义的 be sparked by their desire 被灵活翻译成了表示主动意义的“就是希望”。

2. 倒译法　倒译法，即将原文被动句中谓语后面的宾语倒译成汉语中的主语，而原文的主语则相应变成了汉语中的宾语。

【例 7】The special interest of coastal States in the conservation and management of fisheries in adjacent waters was first recognized in the 1958 Convention on Fishing and Conservation of the Living Resources of the High Seas.

【译文】1958 年《捕鱼与养护公海生物资源公约》最早承认了沿海国家对其相邻水域进行养护和渔业管理的特殊利益。

【例 8】This limit was incorporated in the Santiago Declaration of 1952 and reaffirmed by other Latin American States joining the three in the Montevideo and Lima Declarations of 1970.

【译文】1952 年的《圣地亚哥宣言》接受了这一界线。1970 年，拉丁美洲的其他国家加入进来，在《蒙得维的亚宣言》和《利马宣言》中再次确认了这一界线。

3. 习惯译法　习惯译法主要应用于一些固定结构的句子，需要根据汉语的表达习惯进行翻译。最常见的固定结构包括如下几种：

（1）It＋be＋动词过去分词＋that 从句结构，这样的动词包括 ask、assert、feel、prefer、recommend、suggest、said、believe、estimate、announce、point out、claim、declare、expect、hope、consider 等。

在这一结构中，主语是先行代词是 it，翻译成汉语时，根据具体情况，要么添加“有人、人们、大家”等这样的泛指代词作为主语，要么翻译成“据…”字句或汉语的无主句。

【例 9】It is believed that on the basis of this formula developed the traditional three-mile territorial sea limit.

【译文】人们认为传统的三海里领海的主张就建立在这一原则之上。

（2）主语＋be＋动词过去分词＋to do 结构，常用动词和上文（1）中动词的基本重叠，翻译方法也与其相似。

【例 10】The Earth's mantle is estimated to contain between three to six times as much water as in the oceans.

【译文】据估计，地球的地幔层含水量是海洋的 3～6 倍之多。

4. "是…的"结构 凡是着重说明事件是怎样发生或何时何地发生的，或者表示主语是由某材料制成或组成的，这类被动语态可以翻译成汉语的"是…的"结构。

【例 11】Electricity can be generated by water flowing both into and out of a bay.

【译文】水流进和流出海湾时都是可以产生电力的。

英语里被动语态的基本结构很简单，但是在不同的上下文中，翻译成汉语时可能会千变万化。就像刘宓庆先生在《文体与翻译》中所提醒的那样，"把任何一个翻译课题的研究只看作十种、八种翻译程式的运用，囿于程式运作的方寸之地，对翻译实践和翻译研究来说，都是十分有害的"。在翻译被动语态时，必须摆脱原句在形式上的束缚，灵活运用各种手段，使译文不但在内容上忠实于原文，又在表达上通顺自然，符合汉语的表达习惯。

Questions for Discussion:

1. What are the differences between Chinese and English in the use of passive voice?

2. What are the techniques used in converting passive voice in English into Chinese?

Ⅳ. Exercise 1: Sentences in Focus 练习 1：单句翻译

1. Translate the following sentences into Chinese

(1) Before the Convention on the Law of the Sea could address the exploitation of the riches underneath the high seas, navigation rights, economic jurisdiction, or any other pressing matter, it had to face one major and primary issue—the setting of limits.

(2) Traditionally, smaller States and those not possessing large, ocean-going navies or merchant fleets favoured a wide territorial sea in order to protect their coastal waters from infringements by those States that did.

(3) As the work of the Conference progressed, the move towards a 12-

mile territorial sea gained wider and eventually universal acceptance.

(4) The Convention retains for naval and merchant ships the right of "innocent passage" through the territorial seas of a coastal State.

(5) In addition to their right to enforce any law within their territorial seas, coastal States are also empowered to implement certain rights in an area beyond the territorial sea, extending for 24 nautical miles from their shores, for the purpose of preventing certain violations and enforcing police powers.

(6) Perhaps no other issue was considered as vital or presented the negotiators of the Convention on the Law of the Sea with as much difficulty as that of navigational rights.

(7) Countries have generally claimed some part of the seas beyond their shores as part of their territory, as a zone of protection to be patrolled against smugglers, warships and other intruders.

(8) By the late 1960s, a trend to a 12-mile territorial sea had gradually emerged throughout the world, with a great majority of nations claiming sovereignty out to that seaward limit.

(9) At the Third United Nations Conference on the Law of the Sea, the issue of passage through straits placed the major naval Powers on one side and coastal States controlling narrow straits on the other.

(10) The compromise that emerged in the Convention is a new concept that combines the legally accepted provisions of innocent passage through territorial waters and freedom of navigation on the high seas.

(11) To the coastal State falls the right to exploit, develop, manage and conserve all resources to be found in the waters, on the ocean floor and in the subsoil of an area extending 200 miles from its shore.

(12) Phytoplankton, the basic food of fish, is brought up from the deep by currents and ocean streams at their strongest near land, and by the upwelling of cold waters where there are strong offshore winds.

(13) The desire of coastal States to control the fish harvest in adjacent waters was a major driving force behind the creation of the EEZs.

(14) As long-utilized fishing grounds began to show signs of depletion, as long-distance ships came to fish waters local fishermen claimed by tradition, as competition increased, so too did conflict.

(15) Each coastal State is to determine the total allowable catch for each

fish species within its economic zone and is also to estimate its harvest capacity and what it can and cannot itself catch.

2. Translate the following sentences into English

（1）和陆地上的植物一样，浮游植物中有叶绿素捕捉阳光，并利用光合作用把阳光转化成化学能。

（2）只有当环境恶化或资源的枯竭导致经济衰退时，我们似乎才注意到这个问题。

（3）这些化学物质对环境有不利影响。

（4）当时该岛处于外国的统治之下。

（5）联合国发言人表达了对此事的关切，并表示希望能够尽快解决。

（6）中方对南沙群岛及其毗邻海域拥有无可争辩的主权。

（7）这些岛屿一向归中国管辖，这是众所周知的。

（8）多数航道由于泥沙淤积都需要疏浚，穿过滨海地带的航道尤其是这样。

（9）很多渔民不再以捕捞为生，而是到兴旺的建筑业中去找更能赚钱的职业。

（10）渔业活动不能侵犯别国的捕鱼权利。

（11）色彩鲜艳的船只往来于岛屿之间。

（12）这些国家和中国在地理上接近，对于我们发展贸易关系是一个有利的条件。

（13）河流栖息地被称为河岸系统，通常包含茂密的植被和大量的动物种类。

（14）会谈中，两位总统讨论了两国间货物运输的问题。

（15）这个国家单方面决定增加其鲭鱼捕获量，遭到其邻国的严厉批评。

Ⅴ. Exercise 2：Passages in Focus　练习 2：短文翻译

1. Translate the following passages into Chinese

The hot pursuit of a foreign ship may be undertaken when the competent authorities of the coastal State have good reason to believe that the ship has violated the laws and regulations of that State. Such pursuit must be commenced when the foreign ship or one of its boats is within the internal waters or the territorial sea or the contiguous zone of the pursuing State, and may only be continued outside the territorial sea or the contiguous zone if the pursuit has not been interrupted. It is not necessary that, at the time when the foreign ship within the territorial sea or the contiguous zone receives the order to stop, the ship giving the order should likewise be within the territorial sea or the contiguous zone. If the foreign ship is within a contiguous zone, as defined in article 24 of the Convention on the Territorial Sea and the Contiguous Zone, the pursuit may only be undertaken if there has been a violation of the rights for the protection of which the zone was established.

The right of hot pursuit ceases as soon as the ship pursued enters the territorial sea of its own country or of a third State. Hot pursuit is not deemed to have begun unless the pursuing ship has satisfied itself by such practicable means as may be available that the ship pursued or one of its boats or other craft working as a team and using the ship pursued as a mother ship are within the limits of the territorial sea, or as the case may be within the contiguous zone. The pursuit may only be commenced after a visual or auditory signal to stop has been given at a distance which enables it to be seen or heard by the foreign ship.

The right of hot pursuit may be exercised only by warships or military aircraft, or other ships or aircraft on government service specially authorized to that effect. Where hot pursuit is effected by an aircraft:

(a) The provisions of paragraph 1 to 3 of this article shall apply mutatis mutandis;

(b) The aircraft giving the order to stop must itself actively pursue the ship until a ship or aircraft of the coastal State, summoned by the aircraft, arrives to take over the pursuit, unless the aircraft is itself able to arrest the ship. It does not suffice to justify an arrest on the high seas that the ship was merely sighted by the aircraft as an offender or suspected offender, if it was not both ordered to stop and pursued by the aircraft itself or other aircraft or ships which continue the pursuit without interruption.

The release of a ship arrested within the jurisdiction of a State and escorted to a port of that State for the purposes of an enquiry before the competent authorities may not be claimed solely on the ground that the ship, in the course of its voyage, was escorted across a portion of the high seas, if the circumstances rendered this necessary.

Where a ship has been stopped or arrested on the high seas in circumstances which do not justify the exercise of the right of hot pursuit, it shall be compensated for any loss or damage that may have been thereby sustained.

2. Translate the following passages into English

第一条：为行使中华人民共和国对领海的主权和对毗连区的管制权，维护国家安全和海洋权益，制定本法。

第二条：中华人民共和国领海为邻接中华人民共和国陆地领土和内水的一带海域。

中华人民共和国的陆地领土包括中华人民共和国大陆及其沿海岛屿、台湾及其包括钓鱼岛在内的附属各岛、澎湖列岛、东沙群岛、西沙群岛、中沙群岛、南沙群岛以及其他一切属于中华人民共和国的岛屿。

中华人民共和国领海基线向陆地一侧的水域为中华人民共和国的内水。

第三条：中华人民共和国领海的宽度从领海基线量起为十二海里。

中华人民共和国领海基线采用直线基线法划定，由各相邻基点之间的直线连线组成。

中华人民共和国领海的外部界限为一条其每一点与领海基线的最近点距离等于十二海里的线。

第四条：中华人民共和国毗连区为领海以外邻接领海的一带海域。毗连区的宽度为十二海里。

中华人民共和国毗连区的外部界限为一条其每一点与领海基线的最近点距离等于二十四海里的线。

第五条：中华人民共和国对领海的主权及于领海上空、领海的海床及底土。

第六条：外国非军用船舶，享有依法无害通过中华人民共和国领海的权利。

外国军用船舶进入中华人民共和国领海，须经中华人民共和国政府批准。

第七条：外国潜水艇和其他潜水器通过中华人民共和国领海，必须在海面航行，并展示其旗帜。

第八条：外国船舶通过中华人民共和国领海，必须遵守中华人民共和国法律、法规，不得损害中华人民共和国的和平、安全和良好秩序。

外国核动力船舶和载运核物质、有毒物质或者其他危险物质的船舶通过中华人民共和国领海，必须持有有关证书，并采取特别预防措施。

中华人民共和国政府有权采取一切必要措施，以防止和制止对领海的非无害通过。

外国船舶违反中华人民共和国法律、法规的，由中华人民共和国有关机关依法处理。

第九条：为维护航行安全和其他特殊需要，中华人民共和国政府可以要求通过中华人民共和国领海的外国船舶使用指定的航道或者依照规定的分道通航制航行，具体办法由中华人民共和国政府或者其有关主管部门公布。

第十条：外国军用船舶或者用于非商业目的的外国政府船舶在通过中华人民共和国领海时，违反中华人民共和国法律、法规的，中华人民共和国有关主管机关有权令其立即离开领海，对所造成的损失或者损害，船旗国应当负国际责任。

Unit Nine　Marine Engineering

第九单元　海洋工程

Ⅰ. Warm-up for Theme-related Words/Expressions　词汇预习

acid gas　酸性气（体）
acid rain　酸雨
barrage　*n.* 拦河坝
commercial-scale　商业规模的，商业性的
discount rate　贴现率，折现率
ebb　*n.* 退潮，落潮　*v.*（潮水）退落
electrical grid　输电网络；电气网格
fired power plant　火（力发）电厂
flow velocity　流速
fossil fuel　化石燃料，矿物燃料
global warming　全球（气候）变暖
harness　*n.* 利用，治理，控制
head of water　水头
high（low）tide　满（低）潮，高（低）潮（期）
hydraulic　*a.* 水力的，水压的；用水发动的
hydropower　*n.* 水力发电
hydrostatic head　静水头，静压头，静水压头
impoundment　*n.* 收集；蓄水
low-head hydroelectricity　低水头水力发电
lunar cycle　月（亮）运（行）周期
marsh　*n.* 沼泽；湿地
mechanical power　机械功率，机械动力
megawatt（MW）　*n.* 百万瓦特，兆瓦（特）
mudflat　*n.* 泥滩
MWe　兆瓦电力/水当量米数（abbr＝megawatts of electricity/meters of water equivalent）　兆瓦电力/水当量米数
navigation　*n.* 航行，航海；导航
nuclear plant　核电站，核能发电厂
ocean current　洋流，海流
ocean thermal energy conversion（OTEC）　海洋热能转换系统
offshore wind　离岸风，海上风
powerhouse　*n.* 发电站，发电所
pumped storage　抽水蓄能
renewable energy　可再生能源
renewable source　可再生能源
robust　*n.* 鲁棒（性），稳定（性）
seaboard　*n.* 沿海地带；海岸　*a.* 海

岸的
semi-permeable *a.* 半渗透的
shoreline *n.* 海岸线，海滨线
sluice *n.* 水闸，有闸人工水道
solar power 太阳能
tap *v.* 开发，利用；轻击，轻叩
tidal bay 潮汐湾
tidal energy 潮汐能
tidal flushing 潮水冲洗力
tidal power plant 潮汐发电站
tidal range 潮差，潮位变幅，潮汐变化范围
tidal stream 潮流，潮汐河流
tide mill 潮磨；潮轮
turbine *n.* 涡轮（机）
upstream *a.* 向上游的，逆流而上的 *adv.* 逆流地
waterwheel 水车；吊水机
wave energy 波（浪）能；海浪能
wind energy 风能

Ⅱ. Lead-in & Reading 课文导读

Tidal Energy: Its Exploitation and Utilization

(1) With the oceans covering over 70% of the earth's surface, they are the world's largest collector and retainer of the sun's vast energy—and the largest powerhouse in the world. This energy is continually renewed and is available 24/7. Just a small portion of the energy conveniently stored in the oceans could power the world. Jacques Cousteau said it was equivalent to 16,000 nuclear plants. So how can this source of abundant energy be captured and utilized? For over 60 years several forms of tapping this energy have been researched and implemented, and now with fossil fuels running out and

increasingly expensive, they are more than competitive in costs—and the "fuel" is both free and clean. The various form of ocean energy include OCEAN CURRENTS, OFFSHORE WIND, OCEAN THERMAL ENERGY CONVERSION (OTEC), WAVE ENERGY and TIDAL ENERGY, of which, TIDAL ENERGY is one of commonly-seen energies that has since been put to human use.

What Is Tidal Energy?

(2) Tidal energy, also called tidal power, is a form of hydropower that converts the energy of tides into useful forms of power-mainly electricity. Although not yet widely used, tidal energy has potential for future electricity generation. Tides are more predictable than wind energy and solar power. Among sources of renewable energy, tidal energy has traditionally suffered from relatively high cost and limited availability of sites with sufficiently high tidal ranges or flow velocities, thus constricting its total availability. However, many recent technological developments and improvements, both in design and turbine technology, indicate that the total availability of tidal power may be much higher than previously assumed, and that economic and environmental costs may be brought down to competitive levels. Undersea turbines which produce electricity from the tides are set to become an important source of renewable energy for some developed countries. It is still too early to predict the extent of the impact they may have, but all the signs are that they will play a significant role in the future.

(3) Actually, tidal energy is one of the oldest forms of energy used by humans. Indeed, tide mills, in use on the Spanish, French and British coasts, date back to 787 A. D. Tide mills consisted of a storage pond, filled by the incoming (flood) tide through a sluice and emptied during the outgoing (ebb) tide through a water wheel. The tides turned waterwheels, producing mechanical power to mill grain. We even have one remaining in New York—which worked well into the 20th century.

(4) Tidal power is non-polluting, reliable and predictable. Tidal barrages, undersea tidal turbines—like wind turbines but driven by the sea—and a variety of machines harnessing undersea currents are under development. Unlike wind and waves, tidal currents are entirely predictable.

(5) Tidal energy can be exploited in two ways:

(a) By building semi-permeable barrages across estuaries with a high tidal range.

(b) By harnessing offshore tidal streams.

(6) Barrages allow tidal waters to fill an estuary via sluices and to empty through turbines. Tidal streams can be harnessed using offshore underwater devices similar to wind turbines.

(7) Most modern tidal concepts employ a dam approach with hydraulic turbines. A drawback of tidal power is its low capacity factor, and it misses peak demand times because of 12.5 hr cycle of the tides. The total world potential for ocean tidal power has been estimated at 64,000 MWe[1]. The 25–30 ft tidal variations of Passamaquoddy Bay (Bay of Fundy)[2] have the potential of between 800 to 14,000 MWe.

Where Are Good Areas for Exploiting Tidal Energy?

(8) Traditional tidal electricity generation involves the construction of a barrage across an estuary to block the incoming and outgoing tide. The dam includes a sluice that is opened to allow the tide to flow into the basin; the sluice is then closed, and as the sea level drops, the head of water (elevated water in the basin) using traditional hydropower technology, drives turbines to generate electricity. Barrages can be designed to generate electricity on the ebb side, or flood side, or both.

(9) Tidal range may vary over a wide range (4.5–12.4 m) from site to site. A tidal range of at least 7 m is required for economical operation and for sufficient head of water for the turbines. A 240 MWe facility has operated in France since 1966, 20 MWe in Canada since 1984, and a number of stations in China since 1977, totaling 5 MWe. Tidal energy schemes are characterized by low capacity factors, usually in the range of 20%–35%.

(10) The waters off the Pacific Northwest are ideal for tapping into an ocean of power using newly developed undersea turbines. The tides along the Northwest coast fluctuate dramatically, as much as 12 feet a day. The coasts of Alaska, British Columbia and Washington, in particular, have exceptional energy-producing potential. On the Atlantic seaboard, Marine is also an excellent candidate. The undersea environment is hostile so the machinery will

have to be robust.

(11) Currently, although the technology required to harness tidal energy is well established, tidal power is expensive, and there is only one major tidal generating station in operation. This is a 240 megawatt (1 megawatt = 1 MW=1 million watts) at the mouth of the La Rance river estuary on the northern coast of France (a large coal or nuclear power plant generates about 1,000 MW of electricity). The La Rance generating station has been in operation since 1966 and has been a very reliable source of electricity for France. La Rance was supposed to be one of many tidal power plants in France, until their nuclear program was greatly expanded in the late 1960's. Elsewhere there is a 20 MW experimental facility at Annapolis Royal in Nova Scotia, and a 0.4 MW tidal power plant near Murmansk in Russia. UK has several proposals underway.

(12) Studies have been undertaken to examine the potential of several other tidal power sites worldwide. It has been estimated that a barrage across the Severn River in western England could supply as much as 10% of the country's electricity needs (12 GW). Similarly, several sites in the Bay of Fundy, Cook Inlet in Alaska, and the White Sea in Russia have been found to have the potential to generate large amounts of electricity.

What Is the Impact on the Environment?

(13) Tidal energy is a renewable source of electricity which does not result in the emission of gases responsible for global warming or acid rain associated with fossil fuel generating electricity. Use of tidal energy could also decrease the need for nuclear power, with its associated radiation risks. Changing tidal flows by damming a bay or estuary could, however, result in negative impacts on aquatic and shoreline ecosystems, as well as navigation and recreation.

(14) The few studies that have been undertaken to date to identify the environmental impacts of a tidal power scheme have determined that each specific site is different and the impacts depend greatly upon local geography. Local tides changed only slightly due to the La Rance barrage, and the environmental impact has been negligible, but this may not be the case for all other sites. It has been estimated that in the Bay of Fundy, tidal power plants

could decrease local tides by 15 cm. This does not seem like much when one considers that natural variations such as winds can change the level of the tides by several metres.

What Are the Costs of Tidal Energy?

(15) Tidal power is a form of low-head hydroelectricity and uses familiar low-head hydroelectric generating equipment, such as has been in use for more than 120 years. The technology required for tidal power is well developed, and the main barrier to increased use of the tides is that of construction costs. There is a high capital cost for a tidal energy project, with possibly a 10-year construction period. Therefore, the electricity cost is very sensitive to the discount rate.

(16) The major factors in determining the cost effectiveness of a tidal power site are the size (length and height) of the barrage required, and the difference in height between high and low tide. These factors can be expressed in what is called a site's "Gibrat" ratio[3]. The Gibrat ratio is the ratio of the length of the barrage in metres to the annual energy production in kilowatt hours (1 kilowatt hour = 1 KWH = 1000 watts used for 1 hour). The smaller the Gibrat site ratio, the more desirable the site. Examples of Gibrat ratios are La Rance at 0.36, Severn at 0.87 and Passamaquoddy in the Bay of Fundy at 0.92.

(17) Offshore tidal power generators use familiar and reliable low-head hydroelectric generating equipment, conventional marine construction techniques, and standard power transmission methods. The placement of the impoundment offshore, rather than using the conventional "barrage" approach, eliminates environmental and economic problems that have prevented the deployment of commercial-scale tidal power plants.

(18) Three projects (Swansea Bay 30 MW, Fifoots Point 30 MW, and North Wales 432 MW) are in development in Wales where tidal ranges are high, renewable source power is a strong public policy priority, and the electricity marketplace gives it a competitive edge.

What Are Some of the Devices for Tidal Energy Conversion?

(19) The technology required to convert tidal energy into electricity is

very similar to the technology used in traditional hydroelectric power plants. The first requirement is a dam or "barrage" across a tidal bay or estuary. Building dams is an expensive process. Therefore, the best tidal sites are those where a bay has a narrow opening, thus reducing the length of dam which is required. At certain points along the dam, gates and turbines are installed. When there is an adequate difference in the elevation of the water on the different sides of the barrage, the gates are opened. This "hydrostatic head" that is created, causes water to flow through the turbines, turning an electric generator to produce electricity.

(20) Electricity can be generated by water flowing both into and out of a bay. As there are two high and two low tides each day, electrical generation from tidal power plants is characterized by periods of maximum generation every twelve hours, with no electricity generation at the six hour mark in between. Alternatively, the turbines can be used as pumps to pump extra water into the basin behind the barrage during periods of low electricity demand. This water can then be released when demand on the system is greatest, thus allowing the tidal plant to function with some of the characteristics of a "pumped storage" hydroelectric facility.

Why Tidal Energy?

(21) The demand for electricity on an electrical grid varies with the time of day. The supply of electricity from a tidal power plant will never match the demand on a system. But, due to the lunar cycle and gravity, tidal currents, although variable, are reliable and predictable and their power can make a valuable contribution to an electrical system which has a variety of sources. Tidal electricity can be used to displace electricity which would otherwise be generated by fossil fuel (coal, oil, natural gas) fired power plants, thus reducing emissions of greenhouse and acid gasses.

Notes:

[1] MWe 兆瓦（电力），英文 megawatts of electricity 的缩略词，是一种表功率的单位，常用来指发电机组在额定情况下单位时间内能发出来的电量。（瓦的定义是：焦耳/秒，兆瓦的定义是：兆焦耳/秒。）

[2] Passamaquoddy Bay (Bay of Fundy) 帕塞马克第湾（芬迪湾）是美国缅因州及加拿大新不伦瑞克省之间的芬迪湾的一个内湾，是加拿大北部边境十大经典夏日旅游目的

地之一。

[3] Gibrat ratio 源自“Gibrat Law”这一术语，是Robert Gibrat在研究了法国制造业1920—1921年的数据后，于1931年首次提出的均衡效果法则，即公司规模与其增长速度之间的动态模型。也就是说，公司增长的速度与公司在观察期初的规模无关。后来学者多把它称为“Gibrat Law”。本文中的Gibrat ratio是指拦河坝的长度（米）与年电能产量（千瓦小时）之比。

Ⅲ. Teaching Focus (9): Marine English Terminology Formation and Translation 翻译技巧（9）：海洋英语术语的构成与翻译

海洋英语术语是海洋学科领域用来表示概念的称谓的集合，是通过语音和文字来表达或限定海洋科学概念的约定性语言符号，是人们在这一特定学科领域用来交流思想的工具。在我国通常被称为名称或科技名词（不同于语法学中名词）。这类术语可以是一个单词，也可以是一个词组，还可是一个合成词，它们都是根据英语构词法的基本构词规则而产生的。了解海洋英语术语的构词规律，不仅可以了解这类术语的词汇特征及来源，还有助于提高其翻译质量。

1. 海洋英语术语常见的构词法

（1）合成法（compounding）。合成法是海洋英语术语形成和扩展最重要的手段。由于海洋学科术语的特性，所合成的词主要是复合名词，其结构一般为“修饰语＋中心词”，拼写方式包括合写式（无连字符）、分写式（有连字符）和短语式。

①合写式：eyewall 眼墙，眼壁；tortoiseshell 玳瑁壳；damselfish 小热带鱼。

②分写式：vessel-assist tug 助船拖轮；water-tube boiler 水管锅炉；pile-driver 打桩船。

③短语式：high tide mark 高潮标志；gill cover 鳃盖；conservation on biodiversity 生物多样保护；bulk carrier 散装货船；whale shark 鲸鲨；donkey engine（船上用的）辅助发动机。

（2）派生法（derivation）。派生法是指利用具有派生力的前缀或后缀与词根构成新词，派生词的前缀和后缀多源自拉丁语、希腊语和法语的词根或词缀。许多词缀的构词能力和表达能力非常强，且语义范围狭窄、意义固定明确。例如：

① anti-（防，抗，反）：anti-fungal 抗真菌的；anti-biotic 抗生的；

antioxidant 抗氧化剂。

②bio-（生物）：bioaccumulation 生物富集；bioremediation 生物治理。

③eco-（生态）：ecosystem 生态系统；ecocatastrophe 生态灾难。

④hydro-（水）：hydrostatic head 静水头，静压头；hydropower 水力发电。

⑤poly-（多的）：polychaete worm 多毛类环虫；polymetallic nodule 多金属结核。

⑥macro-（大）：macro-organism 大型生物；macrospore 大孢子。

⑦micro-（微小）：microelement 微量元素；micro-encapsulated diet 微胶囊饲料。

⑧-acea（纲名，目名）：crustacea 甲壳纲动物；cetacea 鲸目，鲸类。

⑨-graph（记录工具）：thermohygrograph 温度湿度记录仪（器）；thermosalinograph 温盐仪。

⑩-meter（计，仪）：thermometer 温度计；tachometer 测速仪。

（3）混成法（blending）。在海洋英语术语中，从两个（偶尔两个以上）在拼写或读音上比较适合结合在一起的单词中抽出部分构成新词的方法称为混成法。最常见的结合方式是把前一个词的开头部分和后一个词的结尾部分结合在一起，或把两个词的词头结合在一起从而构成新词。混成词兼具两个旧词的词形和词义。例如：

snark←snake shark 蛇鲨；redox←reduction oxidation 氧化作用；sonar←sound navigation ranging 声呐；chunnel←channel tunnel 海峡通道（铁路），水底铁路隧道；biotech←biology technology 生物技术。

（4）缩略法（abbreviation）。缩略法是将组成术语中的每一个单词的首字母抽出拼接成一个由大写字母构成的字符串，或者将较长的英语单词取其首部或者主干构成与原词意义相同的单词，所构成的词叫缩略词。缩略法是现代海洋科技英语中的一种重要的构词手段。缩略词可分为如下两类：

①首字母缩略词：OTEC←ocean thermal energy conversion 海洋热能转换；MPA ← marine protected area 海洋保护区；FPC ← fish protein concentration 浓缩鱼蛋白；SSM←single species management 单一物种管理；CBD← convention on biological diversity 生物多样性公约；TAC ← total allowable catch 总可捕量；SSHS←the Saffir-Simpson Hurricane Scale 萨菲尔-辛普森飓风强度量表；EEZ ← exclusive economic zone 专属经济区；COLREGS←International Regulation for Preventing Collision at Sea 国际海上避碰规则。

②原词的一部分缩略，构词新词：CO←carbon monoxide 一氧化碳；MAX←maximum 最大的，最大限度的；quake←earthquake 地震；fluidics←fluidonics 射流学。

（5）借用法（borrowing）。海洋科技英语中的许多术语都是通过直接吸收外来专有名词构成的，包括人名、地名、植物、机构以及商标等。例如：

tsunami 海啸（日语）；El Niño 厄尔尼诺（西班牙语）；hurricane 飓风（加勒比语）；crustacean 甲壳动物（拉丁语）；kuroshio current 黑潮（日语）；typhoon 台风（汉语）；alima larva 阿利马幼体；Eyemouth cure 埃茅斯熏制法；Koch's bacillus 郭霍杆菌；Kjeldahl method 凯氏定氮法；Manila rope 马尼拉绳；Steller sea lion 斯氏海狮；Lincoln index 林可指数。

（6）转化法（conversion）。"旧词赋新意""普通词汇专业化"等词义转化法也是创造海洋科技英语术语最常用的一种方法。例如：

ebb 衰退→退潮；gear 装置，工具→渔具；nurse 护士→保育虫；hold 控制，保留→船舱；leader 领导者→网墙，导缆器；wing 翅膀，翼→网袖；fry 油炸（食品）→鱼苗；scale 规模，比例→鳞片。

这类来自普通词汇的专业术语语义特点：①从旧词的核心意义引申而来；②词义单一，不易混淆，应用范围仅局限于本专业领域。

2. 海洋英语术语的翻译方法 海洋英语术语，和其他学科专业术语一样，概念准确，语义单一，简明易懂，见词明义。在翻译时，必须遵循其应用领域和行业的术语规范与标准，符合地道的汉语表达习惯。常见的方法有：

（1）直译（literal translation）。海洋英语术语多表现为复合词、词组或派生词，它们是按照词缀法、复合法和转类法等构词规则构成的，在翻译时，可根据构成术语的各个词素的意义依次直译出来。例如：

global warming 全球变暖；ocean food 海洋食品；seaweed 海草；bioprospecting 生物勘探；mariculture 海水养殖；sea turtle 海龟；water contaminant 水污染物；coastal current 沿岸流；hard-shell（ed）shrimp 硬壳虾；fish-liver oil 鱼肝油。

（2）意译（liberal translation）。某些海洋英语术语在目的语中难以找到相对应的或规范的词汇来表达，或按字面翻译不足以表达其专业深层涵义时，可以采用意译法。意译法主要包括"推演""延伸"和"释义"等三种方法。例如：

"migration"一词词典释义为"移动、迁移、移民"，但在海洋英语中，这个词是指某些水生动物由于生活环境影响和生理习性要求，需要有规律性的定期定向的区域性移动，这时常常是指 fish migration（洄游）。因此，下面的术语都不可照搬词典释义，应进行意译：spawning migration 产卵洄游；breeding migration 生殖洄游；overwintering migration 越冬洄游；migration instinct 洄游本能；migration pattern 洄游模式。

“high seas”是指一国的专属经济区、领海、内水或群岛国的群岛水域以外的全部海域，是对所有国家，不论其为沿海国或内陆国都开放的水域，即“公海”。如果把该术语中的“high”直译为“高的”，就让人不知所云了，因此，在特定的语境中，“high”可引申为“公共的，共有的”。相同意义的表达还包括：the open sea，blue waters，international waters。例如：

Law of the Open Seas 公海法；high sea fishery 公海渔业；freedom of the open sea 公海自由。

有时，如果直译不符合专业术语表达规范，或者译入语读者不明其义，如果加注又使译文显啰唆，可采用释义法进行意译。例如：larval stage 幼体期（鱼类和水生无脊椎动物个体开始胚后发育、从内源营养转向外源营养的发育阶段）；adult stage 成体期（鱼类和水生无脊椎动物个体初次性成熟到衰老死亡的阶段）。

（3）音译（transliteration）。当语际转换出现目的语空白时，海洋英语术语常按其发音译成相对应的汉语，这也是海洋英语术语翻译中常用的一种方法。音译时，必须考虑译名的规范化、大众化和通用性。适合音译的术语主要包括人名、地名、物名、商品、商标、新理论和新概念等。例如：

sonar 声呐；gene 基因；sardine 沙丁鱼；conger 康吉鳗；El Niño 厄尔尼诺；La Niña 拉尼娜；robust 鲁棒；shark 鲨鱼；typhoon 台风；salmon 三文鱼；dolastatin 多拉司他汀（抗肿瘤药）；furan 呋喃：pasteurizer 巴氏灭菌器。

（4）形译（pictographic translation）。形译可分为三种：

①根据实物的形状，用表达其形象的汉语词汇进行翻译（也称象译）。例如：V-slot 三角形槽；Y-mid-oceanic ridge 叉形大洋中脊。

②根据原语的字形用在意义上等值的译语替换。例如：Z-twist 左捻，即纱线的捻向从左下角倾向右上角；S-twist 右捻；B-cut 单脚剪裁，即沿网结外缘剪断一根目脚的剪裁工艺；NT-joining 纵横缝，即网衣纵向边缘和横向边缘间的缝合。

③如果有些字母在形译时找不到对等的译语，也可保留原字母不译，例如：K value K 值，helper T-cell 辅助 T 细胞。或在该字母后加一个“形”字，例如：U-turn（车、船）U 形转弯（掉头），C network C 形网络。这类形译法比前述两种更为普遍。

（5）混译（mixed translation）。在翻译海洋科技英语术语时，音译和意译常会同时兼而采之，即音译之后加上一个表示类别的词，或者前一部分音译，后一部分意译。这种译法既能取原名之音，又能暗示读者事物的类别、属性等范畴词，读者比较容易接受。例如：

sardine 沙丁鱼；shark 鲨鱼；El Niño 厄尔尼诺现象；La Niña 拉尼娜现象；sonar 声呐装置（系统）；valve-guide 阀导；Welsh rarebit 威尔士干酪；the Saffir-Simpson Hurricane scale 莎菲尔-辛普森飓风强度量表；the Glomar Challenger 格洛马尔挑战者（深海钻井船）；Oyashio current 千岛群岛寒流。

（6）零译（zero translation）。在现代海洋科技英语术语中，有些因为过于冗长，直接用缩略语字母比意译更简洁；有些还没有找到合适的目的语术语与之对应，这时，对它们除根据其原组成部分进行意译外，还可直接照搬原文，直接采用英文缩略词。例如：

UNCLOS 联合国海洋法公约 ；EEZ 专属经济区；MPA 海洋保护区；IUCN 世界自然保护联盟；NOAA 国家海洋大气管理署；immobilized DNA 固相 DNA（脱氧核糖核酸）；MSY 最大持续渔获量；MEY 最大经济渔获量；TAC 总允许渔获量；PM emissions PM（颗粒物）排放；PCBs 多氯联苯；IUU fishing 非法、不管制和不报告捕捞。

Questions for Discussion:

1. How many means of terminology formation are there commonly in Marine English? Try to describe the characteristics of the marine English terminologies created.

2. Which methods and techniques are mainly used in the translation of the marine English terminologies? Give some Examples.

Ⅳ. Exercise 1: Sentences in Focus　练习 1：单句翻译

1. Translate the following sentences into Chinese

(1) So how can this source of abundant energy be captured and utilized? For over 60 years several forms of tapping this energy have been researched and implemented, and now with fossil fuels running out and increasingly expensive, they are more than competitive in costs—and the "fuel" is both free and clean.

(2) Among sources of renewable energy, tidal energy has traditionally suffered from relatively high cost and limited availability of sites with

sufficiently high tidal ranges or flow velocities, thus constricting its total availability.

(3) However, many recent technological developments and improvements, both in design and turbine technology, indicate that the total availability of tidal power may be much higher than previously assumed, and that economic and environmental costs may be brought down to competitive levels.

(4) Tide mills consisted of a storage pond, filled by the incoming (flood) tide through a sluice and emptied during the outgoing (ebb) tide through a water wheel.

(5) A drawback of tidal power is its low capacity factor, and it misses peak demand times because of 12. 5 hr cycle of the tides.

(6) The dam includes a sluice that is opened to allow the tide to flow into the basin; the sluice is then closed, and as the sea level drops, the head of water (elevated water in the basin) using traditional hydropower technology, drives turbines to generate electricity.

(7) Tidal range may vary over a wide range (4. 5 – 12. 4 m) from site to site. A tidal range of at least 7 m is required for economical operation and for sufficient head of water for the turbines.

(8) La Rance was supposed to be one of many tidal power plants in France, until their nuclear program was greatly expanded in the late 1960's.

(9) Tidal energy is a renewable source of electricity which does not result in the emission of gases responsible for global warming or acid rain associated with fossil fuel generating electricity.

(10) Changing tidal flows by damming a bay or estuary could, however, result in negative impacts on aquatic and shoreline ecosystems, as well as navigation and recreation.

(11) The few studies that have been undertaken to date to identify the environmental impacts of a tidal power scheme have determined that each specific site is different and the impacts depend greatly upon local geography.

(12) The major factors in determining the cost effectiveness of a tidal power site are the size (length and height) of the barrage required, and the difference in height between high and low tide. These factors can be expressed in what is called a site's "Gibrat" ratio.

(13) The placement of the impoundment offshore, rather than using the conventional "barrage" approach, eliminates environmental and economic problems that have prevented the deployment of commercial-scale tidal power plants.

(14) As there are two high and two low tides each day, electrical generation from tidal power plants is characterized by periods of maximum generation every twelve hours, with no electricity generation at the six hour mark in between.

(15) But, due to the lunar cycle and gravity, tidal currents, although variable, are reliable and predictable and their power can make a valuable contribution to an electrical system which has a variety of sources.

2. Translate the following sentences into English

(1) 净现值和盈利指标都应按其各自的贴现率进行表述。

(2) 上海市防汛指挥部发言人张震宇说，上海正在考虑在长江河口建造一道防洪闸门。

(3) 全球变暖将会导致气候急骤变化，包括更加频繁的洪水、热浪和干旱。

(4) 土耳其计划利用底格里斯河和幼发拉底河的河水开发大型水电项目。

(5) 水库蓄水后，地质环境的改变必将带来新的环境问题

(6) 在天球里，太阳的活动周期也反映在月亮的盈亏圆缺和相对应的潮起潮落的月运周期中。

(7) 这些数据模型是根据天气、洋流、美国海军和美国国家海洋和大气管理局（NOAA）提供的数据制成的。

(8) 海洋热能转换（OTEC）过程利用了旋转涡轮中大体积的水体的温差，来产生连续不断的再生能源。

(9) 尽管离岸涡轮机的建设和维护成本通常比陆基涡轮机要高，但由于很多大城市坐落于海岸之上，海上风电场可得益于更低的传输成本。

(10) 风能是近几年呼声最高的可再生资源——其中一个原因就在于，支持者们认为风能可以降低全球变暖的威胁。

(11) 但这种水闸门从未在一个如此大的大坝上试验过，世界上也没有任何地方有在这个尺度上成功处理泥沙堆积问题的先例。

(12) 海洋能是一种有利于环保可再生的清洁新能源，包括潮汐能、海流能、波浪能、海水盐差能和海水温差能。

(13) 因此，当我们试图刺激 2009 年的经济时，我们应当认识到一个强健

的信用行业内在的风险与优势。

（14）近年来，随着新型给水管材的应用和装修的兴起，常出现高层住宅最高几层用水水压不足的问题。

（15）它在对云层、降水、冰雹和雷暴的气象探测以及在飞机和船舰的导航中，起着日益重要的作用。

Ⅴ. Exercise 2：Passages in Focus　练习 2：短文翻译

1. Translate the following passages into Chinese

The Fundy Ocean Research Center for Energy（FORCE）reached a major milestone in 2014：the installation of the underwater power cables. The four cables laid along the sea floor of the Minas Passage give FORCE the largest power in the world. With a combined length of 11 kilometres，the four 34.5 kV cables have a total capacity of 64 megawatts，equivalent to the power needs of 20，000 homes at peak tidal flows. This subsea infrastructure will allow small turbine arrays to connect to the electricity grid.

As well，in 2014，all four FORCE developers received approval through the developmental feed-in tariff program for a total of 17.5 megawatts of electricity：Minas Energy，4 megawatts（MW）；Black Rock Tidal Power，5 MW；Atlantis Operations Canada，4.5 MW；and Cape Sharp Tidal Venture，4 MW. This approval allows the developers to enter into a 15-year power purchase agreement with Nova Scotia Power. Both Open Hydro and Black Rock Tidal Power have opened offices in Nova Scotia and begun hiring staff. Open Hydro component purchasing and fabrication is well underway.

FORCE has completed construction of the mini-lander，the first underwater platform created for the Fundy Advanced Sensor Technology（FAST）platform program，designed to measure the marine environment in real time via cable connection. FAST responds to an emerging global need to identify and validate suitable tidal sites as well as monitor the environmental conditions at existing sites，which requires the technology and methods to gather data under the extreme conditions of high flow tidal races.

FORCE continues its environmental effects monitoring program and has finalized it's second report covering studies for the 2011 – 2013 period. If in-stream tidal technology is to grow to a larger，commercial scale project， development must happen safely. As turbines are deployed in the Minas Passage，FORCE is committed to understanding what effects they have on the environment，and reporting those effects to the public.

2. Translate the following passages into English

大多数人并没有意识到海洋是一个可为人类所利用的巨大的能源储藏库。利用海洋能源是几项大胆设想中的目标，有助于满足 21 世纪人类对能源的需求。

自古以来，风车就被用来控制潮汐能，即在潮水正常涨落中所蕴含的巨大能量。现代设计要求在潮差至少 3 米的地区，修建横跨狭窄海湾和河口的大型屏障。随着高潮而流进的海水被锁在屏障后面，低潮时，便开闸放水。这种流水，就像河流大坝上的水力发电厂一样，通过驱动涡轮机来发电。这样，潮汐中蕴含的机械能被用来获取电能。一个这样的发电厂从 1966 年以来，就一直在法国西北部的布列塔尼（Brittany）运营，并实验性地修建了几套其他的设施。人们期待在英国西北部的塞汶河、加拿大东部的芬迪湾以及其他合适的地区兴建几项巨大工程。

运用潮汐能无污染，且效率高，但随之而来的潮汐模式的改变可能会对周边的环境造成巨大的破坏。河口丰富的沼泽地和淤泥滩可能遭到破坏或被摧毁。正常的潮水冲刷受到限制，因而来自其他源头的污染物往往会在上游堆积。同时，河水流向会被改变，增加了内陆洪灾的危险。另一方面，上游建造的人工潟湖可用于消遣娱乐，正如法国的潮汐能电厂。

Unit Ten　Maritime Transport

第十单元　航海运输

Ⅰ. Warm-up for Theme-related Words/Expressions　词汇预习

admiral　*n.* 海军上将；舰队司令；将军；元帅

arc　*n.* 弧，弧线（复数为arcs）；弧形物

artery　*n.* 干线，要道；[解剖] 动脉

backhaul　*n.* 回程，返程；回程运费；回运

bauxite　*n.* 铝土矿，铝矾土

break-bulk cargo　杂货；杂货运输

bulk cargo　散货；散货运输

containerization　*n.* 货柜运输；集装箱运输

contemporarily　*adv.* 同时期地，同时代地

converge　*v.* 聚集；会于一点；接轨

cruise　*n./v.*（军舰等）巡航；巡游；漫游

deadweight　*n.* 重负，重担；吨位

deviation　*n.* [航] 偏航；背离；离经叛道行为

draft　*n.* 草稿；拟量

dredge　*v.* 疏浚；挖掘；使显露；采捞

drum　*n.* 桶；鼓

enclave　*n.* 被包围的领土（区域）；飞地（指在甲国境内的隶属乙国的一块领土）

ferry　*n.* 渡船；渡口；摆渡　*v.* 航海；渡运；摆渡

fluvial　*a.* 河的，河流的，河中的

hinterland　*n.* 腹地；内地；穷乡僻壤

hub　*n.* 轮轴；枢纽；焦点；插孔；[计] 集线器

hydrographic　*a.* 与水道测量有关的，与水文地理有关的

integration　*n.* 整合，一体化；结合；混合；集成

intensification　*n.* 强化；激烈化，增强明暗度

intermediate　*a.* 中间的，中级的　*v.* 调解；干涉　*n.* 中间分子，中间人

loadline　*n.* 载重线

loop　*n.* 回路；圈，环　*v.*（使）成环，（使）成圈

maneuverability　*n.* 可操作性，机动性；可控性

massification　*n.* 大众化；群众化

pendulum　*n.* 摆，钟摆；摇摆不定的事态（或局面）

physiography *n.* 地文学，地相学

port call 沿途到港停靠

reciprocity *n.* 互惠；相互作用；相互性；互给

refinery *n.* 精炼设备；提炼厂；冶炼厂

sailship *n.* 帆船

seaborne trade 海运贸易

seasonality *n.* 季节性

transoceanic *a.* 在海洋彼岸的，横越海洋的

transshipment *n.* 转载

uniform *a.* 一样的；规格一致的；始终如一的 *v.*（使）规格一致；（使）均一

vector *n.* 航线；矢量，向量 *v.* 用无线电导航；为…导航

vertically *adv.* 垂直地；直立地；陡峭地

Ⅱ. Lead-in & Reading 课文导读

Maritime Routes and Traffic

(1) From its modest origins as Egyptian coastal and river sailships around 3200 BC, maritime transportation has always been the dominant support of global trade. By 1200 BC Egyptian ships traded as far as Sumatra, representing one of the longest maritime route of that time. By the 10th century, Chinese merchants frequented the South China Sea and the Indian Ocean, establishing regional trade networks. In the early 15th century, Admiral Zheng He led a large fleet of 317 vessels manned by 28,000 crewmen to conduct seven major expeditions, one of which reached the east African coast. However, China's

attempt at asserting a regional maritime dominance was short lived and such expeditions were not permitted to continue. European colonial powers, mainly Spain, Portugal, England, the Netherland and France, would be the first to establish a true global maritime trade network from the 16th century. Most of the maritime shipping activity focused around the Mediterranean, the northern Indian Ocean, Pacific Asia and the North Atlantic, including the Caribbean. Thus, access to trade commodities remains historically and contemporarily the main driver in the setting of maritime networks.

(2) With the development of the steam engine in the mid 19th century, trade networks expanded considerably as ships were no longer subject to dominant wind patterns. Accordingly and in conjunction with the opening of the Suez Canal, the second half of the 19th century will see an intensification of maritime trade to and across the Pacific. In the 20th century, maritime transport grew exponentially as changes in international trade and seaborne trade became interrelated. Maritime transportation, like all transportation, is a derived demand that exists to support trade relations. These trade relations are also influenced by the existing maritime shipping capacity. There is thus a level of reciprocity between trade and maritime shipping capabilities. As of 2006, seaborne trade accounted for 89.6% of global trade in terms of volume and 70.1% in terms of value. Maritime shipping is one of the most globalized industries in terms of ownership and operations.

(3) Maritime transportation, similar to land and air modes, operates on its own space, which is at the same time geographical by its physical attributes, strategic by its control and commercial by its usage. While geographical considerations tend to be constant in time (with the exception of the seasonality of weather patterns), strategic and especially commercial considerations are much more dynamic. The physiography of maritime transportation is composed of two major elements, which are rivers and oceans. Although they are connected, each represents a specific domain of maritime circulation. The notion of maritime transportation rests on the existence of regular itineraries, better known as maritime routes.

(4) Maritime routes are corridors of a few kilometers in width trying to avoid the discontinuities of land transport by linking ports, the main elements of the maritime / land interface. Maritime routes are a function of obligatory

points of passage, which are strategic places, of physical constraints (coasts, winds, marine currents, depth, reefs, ice) and of political borders. As a result, maritime routes draw arcs on the earth water surface as intercontinental maritime transportation tries to follow the great circle distance. Maritime routes are linking maritime ranges representing main commercial areas between and within which maritime shipping services are established.

(5) The most recent technological transformations affecting water transport have focused on modifying water channels (such as dredging port channels to higher depths), on increasing the size, the automation and the specialization of vessels (e. g. container ships, tanker, bulk carrier) and developing massive port terminal facilities to support the technical requirements of maritime transportation. These transformations partially explain the development of a maritime traffic that has been adapting to increasing energy demand (mainly fossil fuels), the movements of raw materials, the location of major grain markets and not least to the growth of the trade of parts and finished goods. Yet, this process is not uniform and various levels of connectivity to global shipping networks are being observed. This massification of transport into regular flows over long distances is not without consequences when accidents affecting oil tankers can lead to major ecological disasters (e. g. Amoco Cadiz[1], Exxon Valdez[2]).

(6) Fluvial transportation, even if slow and inflexible, offers a high capacity and a continuous flow. The fluvial / land interface often relies less on transshipment infrastructures and is thus more permissive for the location of dependent activities. Ports are less relevant to fluvial transportation but fluvial hub centers experience a growing integration with maritime and land transportation, notably with containerization. The degree of integration for fluvial transportation varies from totally isolated distribution systems to well integrated ones. In regions well supplied by hydrographic networks, fluvial transportation can be a privileged mode of shipment between economic activities. In fact, several industrial regions have emerged in along major fluvial axises as this mode was initially an important vector of industrialization. More recently, river-sea navigation is also providing a new dimension to fluvial transportation by establishing a direct interface between fluvial and maritime

systems.

(7) The majority of maritime circulation takes place along coastlines and three continents have limited fluvial trade; Africa, Australia and Asia (with the exception of China). There are however large fluvial waterway systems in North America, Europe and China over which significant fluvial circulation takes place. Fluvial-maritime ships enable to go directly from the fluvial to the oceanic maritime network. Despite regular services on selected fluvial arteries, such as the Yangtze, the potential of waterways for passenger transport remains limited to fluvial tourism (river cruises). Most major maritime infrastructures involve maintaining or modifying waterways to establish more direct routes (navigation channels and canals). This strategy is however very expensive and undertaken only when absolutely necessary. Significant investments have been made in expanding transshipment capacities of ports, which is also very expensive as ports are heavy consumers of space.

(8) Not every region has a direct access to the ocean and maritime transport. As opposed to coastal countries, maritime enclaves (landlocked countries) are such countries that have difficulties to undertake maritime trade since they are not directly part of an oceanic domain of maritime circulation. This requires agreements with neighboring countries to have access to a port facility through a highway, a rail line or through a river. However, being landlocked does not necessarily imply an exclusion from international trade, but substantially higher transport costs which may impair economic development. Further, the concept of being landlocked can be at times relative since a coastal country could be considered as relatively landlocked if its port infrastructures were not sufficient to handle its maritime trade or if its importers or exporters were using a port in a third country. For example, France has significant nautical accessibility, but the main port handling its containerized traffic is Antwerp in Belgium.

(9) Pendulum service involves a set of sequential port calls from at least two maritime ranges, commonly including a transoceanic service and structured as a continuous loop. They are almost exclusively used for container transportation with the purpose of servicing a market by balancing the number of port calls and the frequency of services. The term pendulum refers to the back and forth movements between the maritime ranges.

(10) The importance and configuration of maritime routes has changed with economic development and technical improvements. Among those, containerization changed the configuration of freight routes with innovative services. Prior to containerization, loading or unloading a ship was a very expensive and time consuming task and a cargo ship typically spent more time docked than at sea. With faster and cheaper port operations, pendulum routes have emerged as a dominant configuration of containerized maritime networks.

(11) The main advantage of pendulum services is the ability to call several ports and therefore increase the shipload factor. This sequence of ports tends to be highly flexible in terms of which ports are serviced to maximize the market potential. There is however the risk of empty trips (particularly backhauls) and longer service times between distant port pairs along the route. The first pendulum route was set in 1962 by Sea-Land between the ports of New York (Newark facilities), Los Angeles and Oakland by using the Panama Canal. The return trip also included a stop in San Juan (Puerto Rico). The most extensive pendulum services are known as "round-the-world" routes as major maritime ranges of the world are services along a continuous loop. Another recent trend has been the integration and specialization of several routes with feeder ships converging at major maritime intermediate hubs. This is notably the case for Europe (Mediterranean, North Sea and the Baltic) in light of the negative impacts of deviations from main maritime shipping routes in terms of service length and frequency of port calls.

(12) Maritime transportation is dominantly focused on freight since there are no other effective alternatives to the long distance transportation of large amounts of freight. Before the era of intercontinental air transportation, transcontinental passenger services were assumed by liner passenger ships, dominantly over the North Atlantic. Long distance passenger movements are now a marginal leisure function solely serviced by cruise shipping. Still, several oceanic ferry services are also in operation over short distances, namely in Western Europe (English Channel; Baltic Sea), Japan and Southeast Asia (Indonesia and the Philippines).

(13) Maritime traffic is commonly measured in deadweight tons, which refers to the amount of cargo that can be loaded on an "empty" ship, without exceeding its operational design limits. This limit is often identified as a

loadline, which is the maximal draft of the ship and does not account for the weight of the ship itself but includes fuel and ballast water[3].

(14) Maritime freight is conventionally considered in two main markets: bulk cargo and break-bulk cargo. Bulk cargo refers to freight, both dry and liquid, that is not packaged such as minerals (oil, coal, iron ore, bauxite) and grains. It often requires the use of specialized ships such as oil tankers as well as specialized transshipment and storage facilities. Conventionally, this cargo has a single origin, destination and client and prone to economies of scale. Services tend to be irregular, except for energy trades, and part of vertically integrated production processes (e. g. oil field to port to refinery). The dynamics of the bulk market are mainly attributed to industrialization and economic development creating additional demand for resources and energy. Break-bulk cargo refers to general cargo that has been packaged in some way with the use of bags, boxes, drums and particularly containers. This cargo tends to have numerous origins, destinations and clients. Before containerization, economies of scale were difficult to achieve with break-bulk cargo as the loading and unloading process was very labor and time consuming. The dynamics of the break bulk market are related to manufacturing and consumption.

Notes:

[1] Amoco Cadiz “阿摩科·卡迪兹号”油轮。“阿摩科·卡迪兹号”油轮事件是世界历史上最严重的环境灾难之一。1978年3月，满载伊朗原油的美国超级油轮“阿摩科·卡迪兹号”从波斯湾向荷兰的鹿特丹驶去，3月24日航行至法国布列塔尼海岸触礁沉没，漏出原油22.4万吨，污染了350千米长的海岸带。这次漏油事件，对所污染海岸的整个海洋生物以及海鸟来说，其灾难程度是史无前例的。仅牡蛎就死掉9 000多吨，海鸟死亡2万多吨。海事本身损失1亿多美元，污染的损失及治理费用却达5亿多美元，而给被污染区域的海洋生态环境造成的损失更是难以估量。

[2] Exxon Valdez “埃克森·瓦尔迪兹”号油轮。“埃克森·瓦尔迪兹”号油轮漏油事故是世界上代价最昂贵的海事事故。1989年3月24日，美国埃克森公司的这艘巨型油轮在阿拉斯加州美、加交界的威廉王子湾附近触礁，原油泄出达800多万加仑，在海面上形成一条宽约1千米、长达800千米的漂油带。事故发生地点原来是一个风景如画的地方，盛产鱼类，海豚海豹成群。事故发生后，礁石上沾满一层黑乎乎的油污，不少鱼类死亡，附近海域的水产业遭受很大损失，纯净的生态环境遭受巨大的破坏。这是一起人为事故，船长痛饮伏特加之后昏昏大睡，掌舵的三副未能及时转弯，致使油轮一头撞上一处众所周知的暗礁。

[3] ballast water　压载水。压载水指为控制船舶纵倾、横倾、吃水、稳性或应力而在船上加装的水及其悬浮物。当船舶空载时，其稳定性和吃水无法满足安全要求，需要使用压载水调整船舶的漂浮状态；当船舶不均匀装载（部分货舱装货）时，局部受力过大，需要使用压载水调整受力状态；在风暴状态下则需要增加压载水以增大船体的吃水，提高船舶行驶的平稳性。

Ⅲ. Teaching Focus (10): Inversion of Order　翻译技巧 (10): 词序调整

词序又称次序，指的是句子中各个词语或成分排列的先后顺序。在很多情况下，无论是在英语还是汉语中，某个单词或短语在句子中的语法功能往往取决于它在句子所处的位置。英汉对比研究表明，这两种语言中主要成分的词序大体一致，但也有一些不同之处。差异最大的是定语和状语的位置，除此之外，还有宾语、插入语、并列结构以及一些特殊句型结构位置的差异。在英汉互译中，把原文的某种词序在译文中变换成另一种词序的方法称为词序变换法或词序调整法。下面分类探讨英汉翻译中需要调整词序的几种情况：

1. 定语词序的调整

（1）单词后置定语词序的调整。一般来讲，英语中单个单词做定语放在被修饰的中心词之前，汉译时无须调整词序。但有些英语单词，如以-able 或-ible 结尾的形容词做定语时，要放在被修饰的中心词之后；又如修饰 some、any、no、every、something、nothing、anything 等不定代词的定语要放在这些被修饰的词之后，这些单词后置定语在汉译时则需要提前。

【例 1】This is the latest technique available in oceanography on the subject at present stage.

【译文】这是目前能找到的与该课题相关的海洋学最新技术。

【例 2】Researchers wanted to find someone competent and reliable among the regional ocean governance to help them collect more accurate first-hand materials about the local hydrological situation.

【译文】研究人员想在区域海洋管理机构中找一个既有能力又可靠的人，来帮忙他们收集有关本地水文情况的更准确的一手资料。

（2）两个或两个以上前置定语词序的调整。当两个或两个以上的形容词（或名词）共同修饰一个中心词时，通常中英文都可以放在中心词之前，但是这种前置定语在英汉两种语言中的基本排列规则差异迥然。一般而言，英语的排列顺序是由一般到专有、由范围小的到范围大的、由次要的到重要的、由程度弱的到程度强的。而汉语中的这类前置定语的词序排列习惯恰好相反。所以，翻译时会将原句中多个形容词的排列顺序进行适当的调整。

【例 3】a powerful，industrial，socialist country

【译文】一个社会主义工业强国

具体地分解英语中不同类型的定语形容词的排列顺序，可以大致概述如下：限定词→性质（描绘、看法、评论）形容词→大小、长短、高低（矮）形容词→形状形容词→年龄（新旧）形容词→颜色形容词→国籍（地区、出处等）形容词→材料（物质等）形容词→用途（类别）形容词。在英译汉时，以上词序会根据汉语表达习惯适当调整。

【例 4】an abandoned bamboo fishing cage

【译文】一个竹制的废鱼箱

【例 5】the two small new foreign power-sail fishing vessels

【译文】那两艘新的小型进口机帆渔船

（3）分词结构（现在分词、过去分词）或短语结构的定语词序的调整。当动词现在（过去）分词单独做定语时，通常放在所修饰的中心词之前，如：meet the growing demand for aquatic products（满足日益增长的水产品需求），become an honored expert on oceanography（成为一位受人尊敬的海洋学专家），汉译时无须进行词序调整。但如果是动词的现在（过去）分词短语结构做定语，则通常放在所修饰的中心词之后，意思同定语从句差不多。这类后置分词短语结构在译成汉语时，一般需要调整到中心词之前。

【例 6】The change of hydrological regime with reservoir inundation and resettlement is a major factor affecting terrestrial life in that region.

【译文】水库淹没处理和移民安置导致的水文情势改变，是影响该地区水生生物的一个主要因素。

【例 7】The facilities and equipment used by traditional sea freight are well positioned to develop modern logistics.

【译文】传统的海洋货物运输所使用的设备和船舶对现代物流的发展起到了很好的定位作用。

（4）中心词兼有前置和后置定语词序的调整。英语句子中有的中心词既有前置定语，又有介词短语、分词短语或不定式短语做后置定语，翻译时须将后置定语改译为前置定语，而原有的前置定语则需要视情况而定。

【例 8】The emerging marine industry will inevitably face various problems arising from its growing up process.

【译文】新兴的海洋产业将不可避免地面临其发展过程中出现的各种问题。

【例 9】Another common misunderstanding related to this kind of fish consists in confusing its dormancy period with normal physiological sleep.

【译文】有关这种鱼类的另一个常见的误会是将其休眠期与正常生理性睡眠混淆。

（5）定语从句词序的调整。英语中的定语从句，无论是限制性定语从句还是非限制性定语从句，一般均置于所修饰的中心词之后。在译成汉语时，往往将较短的限制性定语从句译成前置定语，而较长的限制性定语从句和非限制性定语从句则根据汉语的表达习惯，大多置于中心词之后，译成并列分句或独立的句子。本单元着重举例说明将英语定语从句调整为汉语前置定语的情况。

【例 10】It may call the attention of the government departments concerned，like National Bureau of Oceanography，to situations which are likely to endanger the marine ecological environment and sustainability.

【译文】它可以提醒有关政府部门（如国家海洋局）注意到那些足以危及海洋环境及其可持续发展的形势。

【例 11】Usually the estuary area is naturally eutrophic because it is the place where land-derived nutrients are concentrated.

【译文】通常河口地带天然富营养化，因为这里是陆地衍生的养分聚集的地方。

2. 状语词序的调整 在英语句子中，状语可由单个副词、短语、从句担任，用以表示行为动作的时间、地点、方式、手段、原因、结果、目的等。英语状语可以位于句首、句尾、谓语动词的前后等。当译成汉语时，有的可以保持其原有次序，但是在更多的情况下，需要加以调整，以符合汉语的表达习惯。

（1）英语后置状语调整为汉语前置状语。如上所述，英语状语的位置非常灵活。而按照汉语的表达习惯，状语往往前置。因此，在英译汉时，需要将后置状语调整为前置状语。这种调整不仅适用于用作状语的单个副词或一些短语，而且同样适用于部分状语从句。

【例 12】The Blue Revolution of freshwater aquaculture and mariculture is growing exponentially.

【译文】淡水养殖和海水养殖的“蓝色革命”正在呈指数级增长。

【分析】本例调整了单个副词状语的位置。

【例 13】Container station has a very important consequence and function in this transportation system，the slow development of container station will restrict the level of transport containerization and the transport efficiency of container at a certain extent.

【译文】在这一运输体系中，集装箱场站有着非常重要的地位和作用，集装箱场站发展滞后在一定程度上制约着运输的集装箱化水平和集装箱的运输效率。

【分析】本例调整了短语状语的位置。

【例 14】We should develop containerization of refrigerated transportation as far as possible，because developing intermodal transportation will be the only right choice to develop cold chain of our country.

【译文】应全力发展冷藏运输技术，因为冷藏集装箱多式联运是加快发展我国冷藏链的必然选择。

【分析】本例调整了短语状语的位置。

【例 15】Environmentalists shudder at the prospect of an Arctic criss-crossed by commerce in this way although any development would have to abide by international control on marine pollution.

【译文】虽然所有开发活动都必须遵守国际防止海洋污染的规范，环保主义者对于未来将在北极开辟的纵横交错的商业航线仍感到战栗。

【分析】本例调整了状语从句的位置。

（2）一个句子同时出现几个状语时的词序调整。当几个状语出现在同一句子时，英语的排列次序一般是：方式状语，地点状语，时间状语；而汉语则是：时间状语，地点状语，方式状语。因此，在英汉互译时，对各状语的排列顺序进行调整是必然的。

【例 16】Indiana Standard Oil Company was founded by means of merger in America on June 18th，1889.

【译文】印第安纳美孚石油公司于 1889 年 6 月 18 日在美国以合并的方式成立。

（3）一个句子中同时出现两个（或更多的）时间状语或地点状语时的词序调整。英语句子中若同时出现两个（或更多的）时间状语或地点状语，通常是代表较小单位的状语在前，代表较大单位的状语在后。而在汉语中则相反。所以，翻译时也要进行必要的词序调整。

【例 17】At around 09：45 in the morning of 16 March 1978，the oil tanker Amoco Cadiz ran aground on Portsall Rocks，5 km (3 miles) from the coast of Brittany，France after a failure of her steering system.

【译文】1978 年 3 月 16 日上午 9 点 45 分左右，“阿摩科・卡迪兹号”油轮由于操纵装置失灵，在距法国布列塔尼半岛 5 千米（3 英里）处的波尔萨勒暗礁上搁浅。

3. 宾语词序的调整 在英汉两种语言中，动词宾语（包括直接宾语和间接宾语）可能出现在动词前面或动词后面。在可能的情况下，应尽量保持原来的词序，但如果按原文词序翻译不通，或者译文不通畅，则应该考虑重新调整译文顺序。在调整顺序时，可能会有词语的增减。

【例 18】They plan to locate the new fresh water lobster breeding base in a certain lakeside area of the province.

【译文】他们计划把新的淡水龙虾养殖基地定在该省的某个湖畔区域。

【例 19】Scientists know little about tsunami, and now deep-ocean pressure sensors and coastal tide gauges are the only tools available to detect and measure it.

【译文】科学家对海啸知之甚少，而目前深海压力探测器和海岸潮汐探测器是仅有的可以用来探测和测量海啸的工具。

【例 20】They assigned member states different tasks and China had undertaken and completed the genome sequencing of seven large marine organisms, including oysters, large yellow croaker, grouper, cynoglossussemilaevis, kelp, chlamys scallop in shell and *P. vannamei*.

【译文】他们给成员方分配了不同的工作任务。其中，中国承担并完成了包括牡蛎、大黄鱼、石斑鱼、半滑舌鳎、栉孔扇贝、海带和南美白对虾等七种大型海洋生物的全基因组测序工作。

4. 插入语词序的调整 英语中的插入语通常与其他成分在语法上并无十分密切的联系，如将其省略，句子仍然成立。它可以放在句首、句中或句尾。在译成汉语时，有些插入语，尤其是位于句首的插入语，可保持在原文中的词序。然而，在许多情况下，需对其位置进行调整，如将位于句中、句尾的插入语译至句首，或将位于句中的插入语译至句尾等，以使译文符合汉语的词序排列习惯。

【例 21】The dominant issue in global warming, in my opinion, is sea-level change and the question of how fast ice sheets can disintegrate.

【译文】本人认为，全球暖化的主要议题，是海平面的上升，以及冰原会崩裂得有多快的问题。

【例 22】The design and implementation plan of this hatchery go against the aim to breed better fish, if I may say so.

【译文】如果让我说的话，(我觉得) 该孵化场的设计和实施方案不利于培育出更优良的鱼种。

【例 23】That is the tragedy of the commons and, similarly, the tragedy of fisheries resources on high seas.

【译文】这是公地的悲剧，而公海渔业资源的悲剧也类似。

5. 并列成分的词序调整 当句子中有几个成分并列出现时，英、汉语中的排列次序有时差异颇大。英语中并列的词序一般按照逻辑上的轻重、前后、因果或从部分到整体的顺序安排，而汉语通常则将较大、较强、较突出或给人印象较深的成分前置。所以，在翻译时，需要根据不同语言的表达习惯酌情进行词序调整。

【例 24】Successful enterprise, from small and medium-sized enterprises (SMEs) to transnational ones, can recognize and take seriously the value of an information system, and China Ocean Shipping Group Company (COSCO) is no exception.

【译文】从中小型企业到跨国公司，成功的企业都会认识并重视信息系统的价值，中国远洋运输集团公司（中远）也不例外。

【例 25】Ports as land and water transportation hub, plays an important role in land and sea transport connection, transit passengers and cargo transport, collection and distribution of modern logistics.

【译文】港口作为水陆交通运输的枢纽，发挥着连接海陆运输，中转旅客和货物运输，集散现代物流的重要作用。

【例 26】Of that total tropical shrimp trawl fisheries have the highest "discard" rate, accounting for 27 percent of the wastes and 1.8 million tonnes of it.

【译文】热带虾类拖网渔业的"丢弃"率最高，以 180 万吨之重，占总鱼废弃物的 27%。

6. 特殊句型结构词序的调整 由于语法、修辞、句子结构或表达习惯等原因，英语中有一些特殊的句型句式，如倒装句、强调句、被动句和 it 做形式主语的句子等。在这些句子中，句子成分的基本顺序发生了变化。有的是出现所谓的"倒装词序"；有的是句子的形式主语与逻辑主语各有所属；有的是动作的实施和授施的位置调换了。在翻译这些句子时，首先要正确理解原句内容，然后才能根据汉语的习惯，酌情进行词序调整处理。

【例 27】Never before have marine scientific research and marine resources development and utilization in the academies been as discussed warmly as it is at present stage.（否定词前置的倒装句）

【译文】当今，学术界对海洋科学研究和海洋资源开发利用的探讨空前热烈。

【例 28】Only after seeing through the operation mode of the aquaculture industry chain can fish farmers adapt to the new situation，embrace opportunities，and achieve success in new competition.（only 引导的倒装句）

【译文】只有看清水产养殖行业链的运行模式，养鱼户才能适应新的形势，把握机遇，在新的竞争中立于不败之地。

【例 29】Not until small and medium fish farmers have equivalent access to credit and technical assistance can the path to benign competition and sustainable growth begin in the world market of aquatic products.（倒装句）

OR：It is not until small and medium fish farmers have equivalent access to credit and technical assistance that the path to benign competition and sustainable growth can begin in the world market of aquatic products.（强调句）

【译文】只有在中小型养殖户都能获得同样的贷款和技术援助之后，国际水产品市场才有望踏上良性竞争和可持续的发展道路。

【例 30】Hydrographic and bathymetric data have been applied in this article to review the dynamic regime and morphological evolution in recent years in the sandbar area of the Modaomen estuary.（被动句）

【译文】本文应用近几年的水文和地形实测资料，研究了磨刀门河口沙洲区域近期的动力和地貌演变特点。

【例 31】It is biologically reasonable for crocodile ice fishes to reduce their cost of living to increase their chances of surviving in polar coldness.（it 做形式主语的句子）

【译文】鳄冰鱼减少生存所需的能耗以增加在极地严寒环境下存活下来的机会，这一点从生物学角度来看是合情合理的。

词序调整法是一种常见的翻译技巧。在通常情况下，为了忠实于原文，译者会尽可能地保持译文与原文在句子结构和词序方面的一致性。但是，英汉两种语言之间客观存在的差异，又使得适当的词序或句序调整不可避免。如英语句子词序一般是根据语境来安排的，而汉语通常按事件发生的先后顺序排列。英语常常开门见山，先说主题，后做解释；先呈现结果，后说明原因；先得出结论，后附加条件；先做出判断，后列举现象；先表明态度，后描述事实。而汉语在表达以上事理时的惯用顺序恰好相反。此外，语法结构、词语搭配、惯用表达等方面的不同都是译者在翻译实践时，对原文的词序进行适当调整的重要依据。

Questions for Discussion:

1. What do you think are the basic principles for inverting orders of words in Chinese-English translation?

2. Generalize some other useful ways of inverting words orders from your personal translation practice and share them with your classmates.

Ⅳ. Exercise 1: Sentences in Focus 练习 1: 单句翻译

1. Translate the following sentences into Chinese

(1) Thus, access to trade commodities remains historically and contemporarily the main driver in the setting of maritime networks.

(2) With the development of the steam engine in the mid 19th century, trade networks expanded considerably as ships were no longer subject to dominant wind patterns.

(3) Maritime transportation, similar to land and air modes, operates on its own space, which is at the same time geographical by its physical attributes, strategic by its control and commercial by its usage.

(4) As a result, maritime routes draw arcs on the earth water surface as intercontinental maritime transportation always tries to follow the great circle distance.

(5) The most recent technological transformations affecting water transport have focused on modifying water channels (such as dredging port channels to higher depths), on increasing the size, the automation and the specialization of vessels (e. g. container ships, tanker, bulk carrier) and developing massive port terminal facilities to support the technical requirements of maritime transportation.

(6) Ports are less relevant to fluvial transportation but fluvial hub centers experience a growing integration with maritime and land transportation, notably with containerization.

(7) Despite regular services on selected fluvial arteries, such as the Yangtze, the potential of waterways for passenger transport remains limited to fluvial tourism (river cruises).

(8) However, being landlocked does not necessarily imply an exclusion

from international trade, but substantially higher transport costs which may impair economic development.

(9) Further, the concept of being landlocked can be at times relative since a coastal country could be considered as relatively landlocked if its port infrastructures were not sufficient to handle its maritime trade or if its importers or exporters were using a port in a third country.

(10) Pendulum service involves a set of sequential port calls from at least two maritime ranges, commonly including a transoceanic service and structured as a continuous loop.

(11) Prior to containerization, loading or unloading a ship was a very expensive and time consuming task and a cargo ship typically spent more time docked than at sea.

(12) Another recent trend has been the integration and specialization of several routes with feeder ships converging at major maritime intermediate hubs. This is notably the case for Europe (Mediterranean, North Sea and the Baltic) in light of the negative impacts of deviations from main maritime shipping routes in terms of service length and frequency of port calls.

(13) Before the era of intercontinental air transportation, transcontinental passenger services were assumed by liner passenger ships, dominantly over the North Atlantic.

(14) Maritime traffic is commonly measured in deadweight tons, which refers to the amount of cargo that can be loaded on an "empty" ship, without exceeding its operational design limits. This limit is often identified as a loadline, which is the maximal draft of the ship and does not account for the weight of the ship itself but includes fuel and ballast water.

(15) Conventionally, this cargo has a single origin, destination and client and prone to economies of scale. Services tend to be irregular, except for energy trades, and part of vertically integrated production processes (e. g. oil field to port to refinery).

2. Translate the following sentences into English

(1) 在轴辐式运输模式中，完成支线运输任务的多为中小型集装箱船舶，这些船舶能够很好地完成枢纽港与腹地、枢纽港与喂给港之间的货物集散运输任务。

(2) 船员正在等待支线驳船上的中转集装箱转装到本航次上。

(3) 进行优化设计的目的是解决船体内部存储空间及结构自重与船舶灵活性之间的矛盾。

(4) 文章提出了一种计算自动气象站测量结果的标准差、合格率、准确性的方法。

(5) 钟摆式海运服务也有缺点，例如，有空跑的风险，且沿线远距离港口之间的航行时间过长。

(6) 这些发展的结果是急需一个无线回程传输网络系统，以便提高交互操作性、端到端的安全性以及端到端的服务质量。

(7) 溢流疏浚法主要用于疏浚航道中的泥沙，且一般只在退潮时使用。

(8) 平均每年有40艘美国军舰在中国香港停靠，水手们在商店、酒店、餐馆和酒吧花费的美元数以百万计。

(9) 以前，深圳的市场竞争优势之一在于其毗邻香港——一个思想、理念、建议和投资的来源地。

(10) 随着贸易的发展，以及单件杂货和部分散货运输集装箱化率的提高，集装箱货运量呈现持续增长的态势。

(11) 应该采用面向对象技术来进行系统程序设计，因为该技术能减少代码量，使整个系统可靠、高效，所有程序结构清楚明了，便于维护和拓展。

(12) 我们将根据互惠互利和注重实效的原则，采取具体步骤，促进稳定发展。

(13) 远程特快货运列车上配有能迅速装卸货物的设施。它们是整个运输系统不可或缺的一部分，承担着工业中心和海港城市之间货物运输的任务。

(14) 针对数据库云交付而优化的基础架构通过自动化和硬件标准化突显了简单性和效率。

(15) 许多交通工作者和城市政府开始注重优先发展公共交通，以实现城市土地的集约化利用和可持续发展，避免重走西方发达国家以小汽车出行为主的老路。

Ⅴ. Exercise 2：Passages in Focus　练习 2：短文翻译

1. Translate the following passages into Chinese

The global maritime shipping industry is serviced by about 100,000 commercial vessels of more than 100 tons falling into the following four broad categories：

<u>**Passenger vessels**</u> historically played an important role since they were the

only mode available for long distance transportation. In a contemporary setting, passenger vessels can be divided into two categories: passenger ferries, where people are carried across relatively short bodies of water (such as a river or a strait) in a shuttle-type service, and cruise ships, where passengers are taken on vacation trips of various durations, usually over several days. The former tend to be smaller and faster vessels, the latter are usually very large capacity ships having a full range of amenities. In 2012, about 20.3 million passengers were serviced by cruise ships, underlining an industry with much growth potential since it services several seasonal markets where the fleet is redeployed to during the year.

Bulk carriers are ships designed to carry specific commodities, and are differentiated into liquid bulk and dry bulk vessels. They include the largest vessels afloat. The largest tankers, the Ultra Large Crude Carriers (ULCC) are up to 500,000 deadweight tons (dwt), with the more typical size being between 250,000 and 350,000 dwt; the largest dry bulk carriers are around 400,000 dwt, while the more typical size is between 100,000 and 150,000 dwt. The emergence of liquefied natural gas (LNG) technology enabled the maritime trade of natural gas with specialized ships.

General cargo ships are vessels designed to carry non-bulk cargoes. The traditional ships were less than 10,000 dwt, because of extremely slow loading and off-loading. Since the 1960s these vessels have been replaced by container ships because they can be loaded more rapidly and efficiently, permitting a better application of the principle of economies of scale. Like any other ship class, larger containerships require larger drafts with the current largest ships requiring a draft of 15.5 meters.

Roll on-Roll off (RORO) vessels, which are designed to allow cars, trucks and trains to be loaded directly on board. Originally appearing as ferries, these vessels are used on deep-sea trades and are much larger than the typical ferry. The largest are the car carriers that transport vehicles from assembly plants to the main markets. Their capacity is measured in the amount of parking space they are able to offer to the vehicles they carry, mostly measured in lane meters.

The distinctions in vessel types are further differentiated by the kind of services on which they are deployed. Bulk ships tend to operate both on a regular schedule between two ports or on voyage basis to reflect fluctuations in the demand. This demand may be seasonal, as for grain transport, of niche, such as for project cargo (e. g. carrying construction material). General cargo vessels operate on liner services, in which the vessels are employed on a regular scheduled service between fixed ports of call, or as tramp ships, where the vessels have no schedule and move between ports based on cargo availability.

2. Translate the following passages into English

海洋运输以散货为主，但杂货的份额正稳步增长，这种趋势主要归因于运输集装箱化。与其他运输模式相比，传统的海洋运输面临两大缺陷。首先，它速度缓慢。散装船在海上的平均速度为 15 节（28 千米/时），虽然集装箱船所设计的航行速度在 20 节以上（37 千米/时）。其次，在港口装卸货物会遭遇延误。如果运的是杂货，则处理延误可能需要数天。在短途货物输送或托运人要求快速交货时，这些缺点特别具有约束性。然而，技术改进模糊了散货和杂货之间的区别，因为两者都可用托盘进行单元化装载，而且越来越多的情况下是用集装箱装载。例如，可以将谷物和石油（两者都属散货）装在集装箱内，这种装运方式越来越常见。其结果是，集装箱货运量大幅增长，从 1980 年占非散装货物的 23%到 1990 年的 40%以及 2000 年的 70%。从地理范围来看，海上交通在过去的几十年发展显著，特别是在经过亚欧和跨太平洋贸易增长之后。通过建立大陆间的商业联系，海上运输支撑着一个庞大的交通网。海上运输的优势不在于速度，而在于其运载量和服务的连续性。铁路和公路运输在地理范围和强度方面根本无法支撑这样的交通运输。海洋运输历经了几个主要的技术创新，旨在提高船舶性能或便于它们利用港口设施。这些创新包括：

规模。上世纪，船舶的数量和平均尺寸都有增长。尺寸标志着船的类型和容量。船的大小每增加一倍，其容量增大两倍。规模经济推动着大尺寸船舶提供运输服务。对船东来说，大船的基本工作原理意味着船员、燃料、靠泊、保险和维修费用的减少。所剩唯一的对船舶大小的限制是港口、港湾和运河的容量是否能容纳它们。

速度。船舶航行的平均时速约为 15 节（1 节=1 海里=1 853 米），即每小时 28 千米。在这种情况下，一艘船一天能行驶约 575 千米。目前的船只行驶速度可达 25 到 30 节，但由于能源需求的原因，商船速度鲜有超过 25 节的。为了应对对速度的要求，动力和发动机技术已从帆船发展到蒸汽机、柴油机，再到燃气涡轮机和核动力机。

船舶专业化。规模经济常与专业化相关联，因为许多船只都是专为运载一种货物而设计。这两个过程在很大程度上改变了海上运输方式。久而久之，船舶越来越专业化，包括杂货船、油轮、运粮船、驳船、运矿船、散货船、液化天然气船、滚装船和集装箱船。

船舶设计。从木船，到有钢骨的木船，到钢船，再到用钢、铝和复合材料制造的船，船舶设计已显著改进。今天的船体是人们最大限度地减少能源消耗和建造成本以及提升安全性的努力的成果。

自动化。各种自动化技术都有可能，包括自卸货船、计算机辅助导航和全球定位系统。总的来说，自动化的结果是仅需较少的船员来操作更大的船舶。

Unit Eleven Marine Economy

第十一单元 海洋经济

Ⅰ. Warm-up for Theme-related Words/Expressions 词汇预习

abate v. 减少；[法] 取消法令 n. 减轻；折扣
abet v. 煽动；教唆；怂恿；助长
address v. 应对；处理
adversity n. 逆境；不幸；灾难
alien a. 外国的；相异的 n. 外星人；外国人 v. 让渡；疏远；离间
alleviate v. 减轻，缓和
aquaculture n. 水产养殖；水产业
be geared to/towards 适应；适合
biodiversity n. 生物多样性
biopolymer n. 生物高聚物
brackish water 半咸水，微咸水
breed v. 产生；优生交配；产仔；繁殖 n. 属；种类；类型；血统
capacity n. 容量；性能；生产能力 a. 充其量的，最大限度的
capture fishery 捕捞渔业
catfish n. 鲶鱼
constraint n. 约束；限制；强制
culture v. 培植，培养 n. 养殖
cyprinid n. 鲤科鱼类 a. 鲤科的
effluent a. 发出的，流出的 n. （注入河里等的）污水；流出的水流
enactment n. 制定，颁布；法律，法规；扮演
entrench v. 用壕沟围绕或保护…；牢固地确立
equity n. 公平；公道 a. 股票的；股市的
finfish n. 长须鲸
fishmeal n. 鱼粉
formulation n. 配方；构想；公式化
fragment v. （使）碎裂 n. 碎片；片段
gear n. 齿轮；排挡 v. 接上；调和；使适应；装上齿轮；用齿轮连接
golden apple snail 福寿螺
hatchery n. （尤指鱼的）孵化场
heighten v. （使）变高；（使）加强
incentive n. 动机；诱因 a. 刺激的；鼓励的
incontrovertible a. 无可辩驳的，不容置疑的
institutionalize v. 使…制度化；使成为
market access 市场准入

mitigate v. (使)缓和;(使)减轻;(使)平息
mollusc (mollusk) n. 〈动〉软体动物
niche n. 合适的位置;有利可图的缺口,商机
nuanced a. 有细微差别的;微妙的
offshoot n. 分枝;支流;衍生物;旁系子孙
pangasiid catfish 华鱼芒
pathogen n. 病菌,病原体
perception n. 知觉;观念;觉察(力)
peril n. 危险;冒险 v. 置…于危险中;危及
pest n. 害虫;讨厌的人或事;有害动植物
polymerase n. 聚合酶
predominance n. 优势;主导或支配的地位
procedure n. 程序;步骤;诉讼程序
revenue n. 收益;财政收入;税收收入
setback n. 挫折;退步;阻碍;逆流
shy (away) from 厌恶;害怕;回避
spectacular a. 惊人的;场面壮观的
incontrovertible a. 无可辩驳的
spur n. 马刺;激励因素 v. 鞭策;激励;推动
tariff n. 关税;关税表 v. 征收关税;定税率
territory n. 领地;版图;领域;[商] 势力范围
tilapia n. 罗非鱼
transgenics n. 转基因
trophic chain 营养链
underline v. 在…下面画线;强调
vaccine n. 疫苗,痘苗 a. 疫苗的,痘苗的
whiteleg shrimp (*Litopenaeus vannamei*) 凡纳滨对虾
yield vt. 屈服;生产;获利 vi. 放弃;退让 n. 产量;收益;屈服;产品
indigenous a. 土生土长的;生来的;本地的
market access 市场准入

Ⅱ. Lead-in & Reading 课文导读

Aquaculture in Marine Economies[1] of Asia-Pacific Region

(1) The Asia-Pacific region contributes the major share to global food fish supply from farming; China continues to be the biggest producer. It and seven other countries in the region (India, Indonesia, Thailand, Viet Nam, Bangladesh, the Philippines and Myanmar) are in the top-ten ranked

aquaculture producers in volume and value. The region has a high rate of food fish consumption, estimated at 29 kg per person per year. To maintain this level for the next three decades would require producing an additional 30 to 40 million tons of fish per year by 2050 to meet the demand from a growing population. It has demonstrated the capacity to do so; during this decade many of the countries have produced more food fish from aquaculture than from capture fisheries, and all six countries (China, India, Indonesia, Thailand, Viet Nam and Bangladesh) that have attained a production level of more than one million tons a year are in the region.

(2) Aquaculture systems and species are diverse in the region, but the bulk of its food fish output comes from a few species groups that include cyprinids, tilapias and catfish. All three comprise freshwater species bred in hatcheries, feeding low in the trophic chain and cultured mostly in pond systems. The culture of marine finfish, raised mostly in small floating cages that are located in protected inshore waters is seen to grow rapidly. Large offshore operations using higher-technology cages have begun and are now further adding to marine fish output; however, for technical reasons they are not expected to become widely adopted. The region remains the biggest producer of marine shrimp, now consisting mostly of whiteleg shrimp (*Litopenaeus vannamei*), a Latin American species introduced towards the end of the 1990s. The production of aquatic plants for food, mostly in China and East Asia, is stable, whereas production of aquatic plants for biopolymer, largely in Southeast Asia, is increasingly driven by a rising world

demand. Mollusc production is generally stable but has decreased in some countries. There are a growing number and volume of niche species.

(3) The three major contributions of aquaculture to national economies are GDP, employment and food security. Large or small, commercial farms contribute to social and economic development by supplying aquatic products for consumption, generating business profits, creating jobs, paying wages and salaries, and providing tax revenues. By creating jobs and providing wages, commercial aquaculture helps alleviate poverty. Because this income can be used to purchase food items, commercial aquaculture can improve food security. A significant contribution of commercial aquaculture to food security is its supply of nutritious aquatic food products. Through employment creation and income generation, commercial aquaculture enables more people, especially those in rural areas whose employment opportunities are generally limited, to share the benefits of growth. Therefore, it contributes to the well-being of citizens by providing intra-society equity. Tax revenues from commercial aquaculture constitute resources for stimulating growth, poverty alleviation and food security.

(4) The structure of the sector in much of the region is characterized by the predominance of small-scale independent farms distributed over wide areas and until recently, largely unorganized. The market is also fragmented. These make the management of its development complicated and underline the importance of a strong progressive governance system. Improvements in the governance mechanism have been felt during the past decade, as indicated by fewer conflicts over resources and effluent discharges to public waters, reduction in crop losses from disease, and fewer non-tariff trade barriers faced by shrimp exports. These are largely the outcome of the sector becoming better regulated by a mix of command and control, market-based and voluntary management measures. Organized farmers adopting better management practices have been the key to this progress. The major driver has been market access; although small-scale, almost all of the farms in the region are geared towards producing part or all of the crop to sell to the neighbourhood, the local market or the world concerns for food safety and quality have heightened, largely driven by a more health and quality-conscious public whose purchasing power is becoming stronger. This has been abetted by

the growth in coverage and influence of the modern retail chains. This pressure to produce safe and healthy products in an environmentally responsible way has come from buyers, regulators, civil society and the mass media, transmitted through trade.

(5) Environmental and social issues persist. At the top of the causes of adverse public perception are the feeding of fish with fish and pollution. Substitutes for fish oil and low-value fish (fed directly or as fishmeal) are being developed and tested to mitigate the first issue. Better water and feeding management are helping lessen the volume and organic content of effluent. The region, as with the rest of the world, has shied from transgenics, but it has made effective use of biotechnology products such as vaccines, and of procedures particularly polymerase chain reaction, for health management.

(6) Happily, the destruction of mangroves is now much less of an issue. The ecological and genetic biodiversity impacts of the introduction and transfer of species for culture across borders remain a deep concern among scientists but still needing incontrovertible proof that it has happened, except with the Mozambique tilapia which was introduced into the Asia-Pacific region in the 1950s and has become a pest in fresh and brackish water pond culture systems almost everywhere in the Pacific and in some areas in Southeast Asia. That said, most other introduced species developed for aquaculture have boosted productivity and profitability of farms and have not shown evidence of adverse impacts on biodiversity or the environment, except for a few ornamental fish and the golden apple snail. A slightly different issue is the positive impact of aquaculture on the conservation of marine species and protection of their marine habitats, mainly the coral reefs. This has to do with the increasing use of hatchery-bred seed of some of the species for the live food fish trade. Complete reliance on hatchery-reared seed would in the future likely abate harvesting methods that are destructive to marine life and reefs.

(7) Shrimp continues to be an important earner of foreign exchange for many countries. Tilapia exports are increasing in volume, and the pangasiid catfishes whose sole supplier to the world market is Viet Nam, has entrenched its primacy in major European markets and the United States of America, despite a number of non-tariff barriers to its trade and adverse publicity. The spectacular growth of catfish aquaculture was not predicted, yet it has yielded

a rich vein of lessons in policy, technology application, farm management, and marketing and trade. The core of the lesson lies in its being a relatively low-value fish that small farmers are able to culture at high intensity with yields that no other culture system has come close to achieving, and then its dominance of the markets for white fish in the west. Likewise, many of the achievements in aquaculture development in the region and the process by which they were achieved, the strategies followed and the tools used can be instructive for policy-makers, programme planners, scientists and technologists, advisers to farmers and students. They have not been easy nor cheap to carry out. The sector faced lots of constraints and adversities and suffered many setbacks, yet it finally figured out the ways and means to overcome them.

(8) The region has done its part in implementing all of the 17 action recommendations put forward in the Bangkok Declaration and Strategy[2]. A broad assessment would rate overall performance at above average. This was largely helped by the presence of active regional indigenous organizations, which jointly or individually, often with the technical and/or financial support of international agencies, implemented programmes inspired by the Declaration.

(9) The best achievements were made in aquatic animal health management, pro-poor livelihood oriented aquaculture, small farmer development and inter-regional cooperation. Environmental responsibility is above average, largely due to the wider adoption of better management practices (BMPs) and codes of conduct (CoCs).

(10) Marketing and trade would be scored above average, not because of the increase in trade flow from the region to the traditional major markets, the European Union (EU), Japan and the United States of America, but because of the higher awareness and adoption of food safety and quality standards. Risk management has shown mixed results. Prevention and mitigation of the impacts of biological risks (mainly from pathogens) have been met well by a systematic regional health programme. The risks to biodiversity are being addressed with a regional genetic and biodiversity programme which was the offshoot of an initiative that assessed the impacts of alien and introduced species. However, market-based insurance to enable especially the small farmers to mitigate or cope with the many and increasingly severe perils to

their crop and farm assets is yet to gain headway. A regional initiative on insurance for small farmers has raised awareness and spurred a few activities, but with little progress so far.

(11) Policy support to sustainable development rates an above average mark, with the widespread formulation, enactment and strengthening of policy and development plans for aquaculture and the enabling regulatory measures. This is region-wide, with the Pacific Island Countries and Territories (PICTs) to the Central Asian Republics (CARs) recently adopting national as well as regional aquaculture development policies and plans. Achievements have not been widespread in the three other basic supports to sustainable aquaculture development—education, research, and information. Manpower development has continued at a steady pace via academic training and specialized short courses. The programmes were geared to improving the culture of specific commodities and strengthening specialized support services such as health management, risk management, breeding, molecular genetics and environmental management. Personnel exchanges between Asia and other regions and within the Asian region proceeded at a steady pace.

(12) The core issue of the vast and diverse aquaculture sector of the region is the sustainability of the small-scale farmers that compose most of it. During this decade, the governing sector has gradually moved from compelling farmers to become responsible to providing them with the incentive to produce more with a higher sense of environmental and social responsibility. This needed a nuanced redirection of policy: BMPs, CoCs and market-based incentives began to be more frequently used for sector management and farmer motivation, whereas legal instruments were kept in the background but firmly enforced when needed. To make this governance framework effective, the capacity to provide technical services and management advice to farmers, and the farmers' capacity—by being trained and organized—to make effective use of these, were improved. This, the most important set of strategic lessons that the region has learned during the past two decades, has been internalized in this decade. It should spread and become institutionalized in all the countries.

Notes:

[1] marine economies　海洋经济。现代海洋经济包括为开发海洋资源和依赖海洋空间

而进行的生产活动，以及直接或间接为开发海洋资源及空间的相关服务性产业活动，主要包括海洋渔业、海洋交通运输业、海洋船舶工业、海盐业、海洋油气业、滨海旅游业、海洋服务业。

[2] Bangkok Declaration and Strategy 曼谷宣言与战略。全名为“Aquaculture Development Beyond 2000：the Bangkok Declaration and Strategy”（《2000年后水产养殖业发展战略》）。2000年2月，新千年世界水产养殖大会在泰国曼谷召开。大会预测了未来社会经济的发展对水产养殖业的需求和水产养殖业发展中面临的挑战，并通过了“曼谷宣言与战略”。宣言主要回顾了过去30年水产养殖对全球经济发展所做出的贡献，并提出了新千年水产养殖的发展方向。

Ⅲ. Teaching Focus (11): Division 翻译技巧（11）：分译

英汉两种语言之间的差异很大。英语重形合（hypotaxis），汉语重意合（parataxis）。英语长句较多，汉语较少使用长句。英语的后置修饰语有时很长，而汉语的修饰语一般前置，不宜过长。英语常常用一个词或短语表达一层意思，汉语一般用包含动词的小句表达一层意思。英文句子注重结构的完整与严密，主干部分通常很短，但名词短语、介词短语、分词短语等却较多，它们如众星捧月般围绕着主干结构进行空间搭架，汇集在一起。而汉语句子往往按照事情发展的先后顺序或语义的逻辑性灵活安排语序，更多地利用组合能力强、节奏感突出的分句。因为英汉语言有这些截然不同的外在表现形式，所以译者必须牢记这些差异，灵活运用不同的翻译技巧与方法，化繁为简、化长为短、化整为零。

在众多的翻译技巧和方法中，分译是改变原文句子结构的重要且有效的变通之法。分译法又叫分句法、拆译法、断句译法等。分译就是在翻译时把原句的某个成分从原来的结构中分离出来，译成一个独立成分、从句、并列分句或完整的句子。分译在英译汉中使用较多，不仅可用于处理较长的复杂句，还可用于英语简单句、短语，甚至单词的翻译。下面将分别讨论分译在各种情形下的应用。

1. 分译在英译汉的应用

（1）单词的分译。英语中有些单词由于其在搭配、词义等方面的特点或在原句中的位置，直译会很尴尬或表意不清。这时就应把这类单词从原句中拆离开来，译成一个分句或单句。这样的译句读起来更通顺，不仅不损伤原意，还更符合汉语的表达习惯。英语中的名词、动词、形容词和副词等都可分译。

【例1】The cost of tuna limits its production.

【译文】金枪鱼的养殖成本高，限制了其批量生产。（分译名词）

【例 2】The designs of offshore cage have been found to vary from each other in different breeding bases.

【译文】已经发现，深水网箱的设计在各个养殖基地各不相同。（分译动词）

【例 3】The technology for cage-culture of tuna was the most identifiable big trouble in implementing this project.

【译文】金枪鱼网箱养殖技术是这个项目执行过程中的大问题，这是大家很容易看出来的。（分译形容词）

【例 4】China's long-term modernization program understandably and necessarily emphasizes the growth of marine economy.

【译文】中国的现代化长期规划强调海洋经济的发展，这一点可以理解，也很有必要。（分译副词）

（2）短语的分译。除了单词外，某些英语短语的意思在直译时也无法将其译透。所以，经常需将这些短语结构译成含有动词的小句或是有主谓结构的句子，来充当整个句子的从句或并列分句部分。英语中的名词短语、分词短语、介词短语词和动词不定式短语等都可分译。

【例 5】The report pointed out the increasingly serious overfishing in that region.

【译文】报告指出，该地区过度捕捞的问题日益严重。（分译名词短语）

【例 6】In 1968 the ecologist Garrett Hardin wrote a paper called "The Tragedy of the Commons", required reading for biology students ever since.

【译文】1968 年生态学家迦勒特哈丁写了一篇题为"公地悲剧"的论文，从此以后每位生物学学生都被要求阅读这篇论文。（分译分词短语）

【例 7】Residents from across the state pull together as one community in a massive effort to maintain the ecological environment of our shoreline.

【译文】全州各地居民团结起来，结成一个大团体，通过众人的努力来维护我们海岸的生态环境。（分译介词短语）

【例 8】We will have to mitigate and adapt to avert the most extreme scenario that scientists predicts to take place in the high seas.

【译文】我们必须进行缓解和适应，从而避免科学家们所预测的最极端情境在公海发生。（分译动词不定式短语）

（3）简单句的分译。有些句子虽然是简单句，但因为修饰成分较多，所以句子显得很长。翻译这类简单句，可以把那些修饰成分拆离出来，单独翻译成为一个个小句、分句或并列句，并根据具体情况在必要的地方适当增译。

【例 9】They were forced to put forward an alternative of separate development for all industries in this so-called homeland, with a promise of absolute priority for marine aquaculture in these areas.

【译文】他们不得不变换一下花样，提出在所谓的家园本土上分别发展各种产业，并且承诺在这些地区，海水养殖也会取得绝对优先发展权。

有的简单句虽不长，但照原句翻译，译文很别扭。尤其是原句中个别单词或短语很难处置，不拆开来译就不通顺或生硬晦涩或容易发生误解。这时需对这种单词或短语进行分译处理。

【例 10】Authorities try in vain to convince the local fishermen to abandon illegal fishing on their own.

【译文】有关部门试图说服当地渔民自觉地放弃违法捕捞行为，但他们失败了。

【例 11】Fishermen in that region might have spoken with understandable pride of their fishery yield.

【译文】该地区的渔民谈到他们的渔业产量时候，也许有自豪感，这是可以理解的。

【例 12】They submitted full specifications of their fishing gear together with term of payment and discount rate.

【译文】他们提供了公司所生产的各种捕捞渔具的详细规格说明，并告知付款条件和折扣率。

（4）复合句的分译。英语复合句主要分为并列复合句和主从复合句两大类。

①并列复合句：有些并列复合句的结构不算很复杂，在分译时，一般直接在并列联词处进行断句处理即可。

【例 13】One camp held that they had transitioned fully to terrestrial life, and that the tree-friendly features of the upper body were just evolutionary baggage handed down from an arboreal ancestor.

【译文】论战的一个阵营认为它们已经转变为陆地生物，其上肢对于树的适应特征只是来自树栖远祖的进化残留。

有的并列复合句则在其中一个并列部分出现二级修饰成分，此时，需要根据句意，在必要的地方进一步拆分，将整句译成更多的句子或分句。

【例 14】That leads to a runaway greenhouse effect and that, in turn, makes Venus awfully hard to hold on to the oceans and seas that served as the incubators for all terrestrial life.

【译文】这就造成了失控的严重温室效应，其反过来又会让海洋很难在金星表面存在，而海洋是孕育陆地上一切生命的摇篮。

【例 15】Elastase activity is measured by elastin—orcein and that is contributed to the alkaline protease activity but there are problems in correlating the results on this activity using different substrates.

【译文】弹性蛋白酶的活性是用弹性蛋白——地衣红做底物测定的。结果表明，弹性蛋白酶的活性受碱性蛋白酶的活性推动，但是，用不同底物来比较酶的活性还存在一定问题。

有些并列复合句内的修饰成分层层叠加，修饰语与被修饰部分的关系很不清晰。对于这样结构过于复杂的复合句，需要先进行严格的语法分析，然后再视情况分译。

【例 16】(a) Utilization of dietary protein and energy by rainbow trout is not (b) as efficient as is generally believed or (c) as it could be, (d) at least as far as total carcass production is concerned and (e) considerable improvement can be achieved (f) provided more attention is drawn to the interrelationships between dietary protein- and energy-yielding nutrients.

【译文】虹鳟对饲料中的蛋白质和能量的利用率不如公认的那样高，即未达到所应达到的利用率，至少从总增肉量来看是这样；如果今后加强对产生蛋白质的营养素与产生能量的营养素间的关系的研究，就会大幅度地提高虹鳟对饲料蛋白和能量的利用率。

【分析】这个并列复合句包含六个从句，前四个从句为一个意群，后两个从句为一个意群，两个意群并列。(b)、(c)、(d) 均为比较状语从句，(e)、(f) 为条件状语从句。可根据意群的并列情况以及从句之间的逻辑关系进行翻译。

②主从复合句：英语主从复合句中的从句类型很多，语法功能也不尽相同。但一般来说，在英译汉时，会将原文的从句译成汉语的分句或独立句或另一种从句，以使译文通顺流畅或层次分明，更合乎汉语习惯。如果从句中出现较长的成分或不易安排的成分还需要进一步拆开来译。

【例 17】Fishmeal is an important animal protein feed that is rich in protein and vitamins and is a major raw material of animal feed.

【译文】鱼粉是畜禽的重要动物性蛋白质饲料，其蛋白质含量高且具有丰富的维生素，是饲料工业主要的动物性饲料原料。

【例 18】It propels the China Marine Surveillance to find solutions to the problems that they never would have considered if not for adversity.

【译文】它驱使中国海监总队找到解决问题的方法，如果不是身处逆境，也许他们永远不会考虑到。

【例 19】Stop building new coal-fired power plants in the river or offshore areas unless they can capture and sequester the carbon they emit.

【译文】停止在河流或近海区域建造新的燃煤发电厂，除非他们可以将工厂排放的碳捕捉并隔离。

有的复合句的主句与从句或主句与修饰语之间的关系并不十分密切，在翻译时需要先进行细致的语法分析，然后按照汉语多用短句的习惯，把从句或短语化为句子，断开来叙述；为使语意连贯，还可适当增加词语。

【例 20】Thus a volcano which forms on a moving plate above a plume will eventually move away from the rising column，which will then melt through at a new location and form another volcano，while the old volcano becomes extinct.

【译文】因此，形成于一个地幔羽之上移动板块上的火山，最后会从上升的地幔柱的位置上移开。然后，地幔柱将在一个新的位置熔穿并形成另一座火山，而老的火山就变成死火山。

【分析】本例中主要需弄清第二个 which 所指代是什么，这是厘清此句逻辑关系的关键。经过分析，可以看出这个 which 指代的应是前面的 the rising column，连接词 then 表明了前后的时间关系。由此，我们可以将这一长句拆分为汉语的两个分句，其表达可能会更加符合汉语的习惯，意思更清楚了然。

总之，英语复合句翻译的情况变化不定，需要在实战中不断积累经验，灵活处理。分译法是英语复合句翻译时的常用方法，尤其是在翻译结构复杂且纠结的长句时。一般情况下，原封不动地按原句的结构翻译这类复合句，译文会让人感到费解。所以必须先对其进行语法分析，找出句子的主语、谓语和宾语，然后找出修饰主语和宾语的定语和定语从句，以及修饰谓语的状语和状语从句。厘清句子的语法层次后，再按汉语的语言习惯调整搭配，恰如其分地将句子断开，分译成多个短句，才能将句子意思表达清楚。

2. 分译在汉译英的应用　分译也可用于汉译英，尤其是在翻译汉语长句时，有时需要将汉语的一个句子拆译成两个或两个以上的英语句子。

【例 21】这次暂停政策于去年 10 月结束，但自那时起，政府只批准了少量的海洋深水钻井。

【译文】That moratorium ended in October. But the government has given very few permits on offshore deep water drilling since then.

【分析】当汉语句子有语气或话题的转折时，通常进行分译。

【例 22】总之，中国主办的亚太水产养殖展览会，不仅有利于进一步改善亚太地区水产品贸易环境，加强国际合作，而且有利于中国在水产业中发展市场经济，增强中国水产品在国际市场的竞争力，这是发展中国水产养殖业的一个非常难得的机会和积极因素。

【译文】In short, the Asia Pacific Aquaculture Exhibition hosted by China will not only help improve the regional aquatic products trade environment and strengthen its international cooperation, but also promote the development of market economy in China's fisheries sector and enhance the competitive capacity of China's aquatic products on the international market. This is indeed a rare opportunity for, as well as a positive factor in, the development of China's aquaculture.

【分析】当汉语长句以一个表示判断或小结的从句结尾时，通常进行分译。

【例 23】在现代信息社会里，交通和通信的发展，逐步缩短了地球空间的距离，公路、铁路、航空、水路（包括河道和航道）和管道运输形成了国家综合交通运输网络，成为现代社会经济生活的重要支柱。

【译文】In modern information society, the development of transportation and communication reduced the terrestrial distance gradually. The highway, the railroad, the aviation, the waterway (including river-ways and sea-routes) and the pipeline transportation have formed the national synthesis transportation network, becomes the mainstay of modern economy life.

【分析】当汉语长句中含有从一般到具体或从具体到一般的过渡时，通常进行分译。

分译法是英汉互译时的常用且有效的方法。分译就是把原句中的某个词或短语译为目标语中的单句或分句，或将原来的一个句子，特别是一个又长又复杂的句子译成目标语的两个或两个以上的句子。分译是为了确保译文在忠实于原文的基础之上，更加清晰易懂，更符合目标语的表达习惯。

Questions for Discussion:

1. How do you define "division"?
2. Illustrate "division" by using some typical examples.

Ⅳ. Exercise 1: Sentences in Focus 练习 1：单句翻译

1. Translate the following sentences into Chinese

(1) It has demonstrated the capacity to do so; during this decade many of the countries have produced more food fish from aquaculture than from capture

fisheries, and all six countries (China, India, Indonesia, Thailand, Viet Nam and Bangladesh) that have attained a production level of more than one million tons a year are in the region.

(2) Large offshore operations using higher-technology cages have begun and are now further adding to marine fish output; however, for technical reasons they are not expected to become widely adopted.

(3) Large or small, commercial farms contribute to social and economic development by supplying aquatic products for consumption, generating business profits, creating jobs, paying wages and salaries, and providing tax revenues.

(4) Improvements in the governance mechanism have been felt during the past decade, as indicated by fewer conflicts over resources and effluent discharges to public waters, reduction in crop losses from disease, and fewer non-tariff trade barriers faced by shrimp exports.

(5) The major driver has been market access; although small-scale, almost all of the farms in the region are geared towards producing part or all of the crop to sell to the neighbourhood, the local market or the world.

(6) The region, as with the rest of the world, has shied from transgenics, but it has made effective use of biotechnology products such as vaccines, and of procedures particularly polymerase chain reaction, for health management.

(7) The ecological and genetic biodiversity impacts of the introduction and transfer of species for culture across borders remain a deep concern among scientists but still needing incontrovertible proof that it has happened, except with the Mozambique tilapia which was introduced into the Asia-Pacific region in the 1950s and has become a pest in fresh and brackish water pond culture systems almost everywhere in the Pacific and in some areas in Southeast Asia.

(8) Tilapia exports are increasing in volume, and the pangasiid catfishes whose sole supplier to the world market is Viet Nam, has entrenched its primacy in major European markets and the United States of America, despite a number of non-tariff barriers to its trade and adverse publicity.

(9) The core of the lesson lies in its being a relatively low-value fish that small farmers are able to culture at high intensity with yields that no other culture system has come close to achieving, and then its dominance of the

markets for white fish in the west.

(10) This was largely helped by the presence of active regional indigenous organizations, which jointly or individually, often with the technical and/or financial support of international agencies, implemented programmes inspired by the Declaration.

(11) Marketing and trade would be scored above average, not because of the increase in trade flow from the region to the traditional major markets, the European Union (EU), Japan and the United States of America, but because of the higher awareness and adoption of food safety and quality standards.

(12) However, market-based insurance to enable especially the small farmers to mitigate or cope with the many and increasingly severe perils to their crop and farm assets is yet to gain headway.

(13) The programmes were geared to improving the culture of specific commodities and strengthening specialized support services such as health management, risk management, breeding, molecular genetics and environmental management.

(14) During this decade, the governing sector has gradually moved from compelling farmers to become responsible to providing them with the incentive to produce more with a higher sense of environmental and social responsibility.

(15) This needed a nuanced redirection of policy: BMPs, CoCs and market-based incentives began to be more frequently used for sector management and farmer motivation, whereas legal instruments were kept in the background but firmly enforced when needed.

2. Translate the following sentences into English

(1)《养鱼经》，由战国时期（公元前475—前221年）的范蠡所著，是世界上最早的养鱼文献。范蠡在公元前473年开始养殖鲤鱼，该书正是其经验之著。

(2) 文章分析了淡水贝类与其栖息环境的关系以及该流域内贝类的常见种和偶见种，还讨论了研究水域内的贝类的分布界限。

(3) 加强水污染防治，对保障人体健康、减轻环境危害极为重要。

(4) 沉降池内的废水中加入了消毒剂，以消灭所有有毒微生物。

(5) 尽管大量的鱼粉被用作牲畜饲料主要是禽类部门的做法，但是水产养殖目前占了世界鱼粉消费量的35%。

(6) 因此，随着水产养殖对鱼粉需求的增加，其与陆地畜牧业争夺有限资源的竞争将会加剧，对鱼粉价格和供应都将造成影响。

(7) 我们需要开展迅速而连续不断的数据和信息采集，以预报、监测和缓解这样的灾害。

(8) 植物转基因已经日渐成为当今分子生物学和植物育种的最重要的方法和技术之一。

(9) 排水技术的研究对港口码头的稳定性分析以及地质灾害的防治具有指导性意义。

(10) 雨水的利用是减轻北京城市防洪压力和缓解水资源危机的重要措施之一。

(11) 生物多样性变化影响着生态系统功能，而生态系统遭到严重破坏也可能最终影响到维持生命机体的生态系统产品和服务。

(12) 这不仅是一种受欢迎的热带观赏鱼，也是一种具有重要商业价值的食用水产鱼类产品，因为它肉嫩味美，蛋白含量高。

(13) 栖息地破碎化给野生动物带来的不良后果仍然是全球生态学家和保护生物学家共同关心的问题。

(14) 任何单位和个人不得强迫渔民、养鱼户或水产养殖的生产和经营组织购买其指定的捕捞渔具或养鱼设施。

(15) 自然保护主义人士早已认识到了森林对维持发展中国家的土著居民的生计方面起到的重要作用。

Ⅴ. Exercise 2：Passages in Focus 练习 2：短文翻译

1. Translate the following passages into Chinese

Aquaculture refers to the production activity of breeding，rearing，and harvesting aquatic animals and plants under controlled conditions in all types of water environments including ponds，rivers，lakes，and the ocean. Researchers and aquaculture producers "farm" freshwater and marine species of fish，shellfish，plants，algae and other organisms that are widely used in a range of food，pharmaceutical，nutritional，and biotechnology fields. They are different kinds of food fish，sport fish，bait fish，ornamental fish，crustaceans，mollusks，seaweed，sea vegetables，and fish eggs，etc.

Aquaculture is not only a method to produce food and other commercial products，but also an effective way to restore habitat and replenish wild stocks，and rebuild populations of threatened and endangered species. For example, stock restoration or "enhancement" is a special form of aquaculture in which hatchery fish and shellfish are released into the wild to rebuild wild populations or coastal habitats such as oyster reefs.

There are two main types of aquaculture—freshwater aquaculture and marine aquaculture. Freshwater aquaculture produces species that are native to rivers, lakes, and streams, including catfish, trout, tilapia, and bass, etc. Freshwater aquaculture takes place primarily in ponds and in on-land, manmade systems such as recirculating aquaculture systems. Marine aquaculture or mariculture refers to the culturing of species that originally live in the ocean, including oysters, clams, mussels, shrimp, salmon, cod, seabass, and seabream, etc. Marine aquaculture can take place in the ocean (that is, in cages, on the seafloor, or suspended in the water column) or in on-land, manmade systems such as ponds or tanks. Recirculating aquaculture systems that reduce, reuse, and recycle water and waste can also support some marine species. As the demand for seafood has increased in the world market, relevant technologies have been increasingly improving and made it possible to grow food in wider coastal marine waters and the open ocean.

In 2012, the total world production of fisheries was 158 million tonnes, of which aquaculture contributed 66.6 million tonnes, about 42%. The growth rate of worldwide aquaculture has been sustained and rapid, averaging about 8% per year for over 30 years, while the take from wild fisheries has been essentially flat for the last decade. The aquaculture market reached $86 billion in 2009. Aquaculture is an especially important economic activity in China. Between 1980 and 1997, the Chinese Bureau of Fisheries reports, aquaculture harvests grew at an annual rate of 16.7%, jumping from 1.9 million tonnes/year to nearly 23 million tonnes/year. In 2005, China accounted for 70% of world production.

Aquaculture is also currently one of the fastest-growing areas of food production in the U.S. As the nation's oceans agency, National Oceanic and Atmospheric Administration (NOAA) and its Office of Aquaculture focus on marine aquaculture, although research and advancement in technology can be more broadly applied. Continued advances in technology and management practices are expanding aquaculture's potential role in producing a variety of species for both stock restoration and commercial purposes. Also, NOAA is committed to supporting an aquaculture industry that is economically, environmentally and socially sustainable. NOAA experts and partners work to understand the environmental effects of aquaculture in different settings and provide best management practices to help reduce the risk of negative impacts.

2. Translate the following passages into English

中国的海洋经济发展势头迅猛。2001年，中国海洋产业的总产值不到1万亿元人民币；2011年，该数字达到4.56万亿元，平均年增长16.7%，显著高于国民生产总值的增长。目前，中国的海洋总产值占国民生产总值的9.7%，使之成为国民经济的一个新的增长引擎。

在各级政府的推动下，中国的海洋经济在过去的十年里已经走上一条快速发展和持续增长的道路。海洋产业已从早期的以海洋捕捞业和海洋盐业为中心的模式，发展成了一个相当完备的产业体系。该体系由交通运输、滨海旅游、海洋石油和天然气以及造船业引领，并受到如海洋电力工业、水资源利用、海洋生物医药与海洋技术教育服务等新兴产业的大力扶持。

中国传统的海洋产业继续发展。其远洋运输能力不断提高，有20多个港口的航运能力超过1亿吨，在货物吞吐量方面连续七年位居世界第一。中国已成为世界上海洋石油和天然气的最大生产国之一——2010年，海洋天然气和石油的生产首次超过5 000万吨石油当量，使之成为一个“海上的大庆油田”。中国的造船能力在各个层面上都有所提升：2011年，船舶完成数量、在制订单数量和新订单数量都居世界第一。海水养殖能力和远洋捕捞能力显著提高，海洋水产品加工与出口能力也一直在持续上升。

由海洋高技术支撑的新兴战略性海洋产业正在以年均20%以上的增长率快速发展，使其成为中国海洋经济改革，甚至是整个中国经济改革的最大亮点。2011年，海水利用产业的附加值约为10亿元人民币，是“十一五规划”初期的两倍；而海洋可再生能源产业的附加值则接近人民币49亿元，是“十一五规划”初期的10倍多。此外，一些新的产业形式，如游船、游艇、休闲渔业、海洋文化和海洋相关的金融和航运业也已加速发展。

从海上风力发电到海水利用，从海洋生物到海洋矿藏，中国正在全方位开发和利用着本国的海洋资源，进一步深化发展本国的海洋企业。与此同时，中国也将从沿海水域走向公海，从海洋走向两极，深入海洋，进一步探入极地地区，充分展现出中国海洋企业发展的力度。

Unit Twelve　Ocean Management

第十二单元　海洋管理

Ⅰ. Warm-up for Theme-related Words/Expressions　词汇预习

ad-hoc　*n.* 点对点；特设；专门
asphyxiate　*v.* （使）窒息
avert　*v.* 转移；防止，避免
baffling　*a.* 令人困惑的，难对付的，难解的
becalm　*v.* 使（船等）停住
beef up　加强，补充；使更大（更好等）
bilateral　*a.* 双边的；双方的；两侧的
bonanza　*n.* （突然的）财源；富矿脉
brew　*v.* 即将发生；酿造；泡，煮；策划
chlorophyll　*n.* 叶绿素
coral reef　珊瑚礁
crack down on/upon　打击；制裁；镇压
cushion　*n.* 垫子；起保护（或缓冲）作用　*v.* 起缓冲作用；使免遭损害；使柔和的事物
cyanobacteria　*n.* 蓝细菌；蓝藻
deteriorate　*v.* （使）恶化
disastrous　*a.* 灾难性的；悲惨的
dissolve　*v.* （使）溶解；液化；消失；终止
dysfunctional　*a.* 功能失调的；不良的
efficiency gains　效率利得；效率收入
exclusive　*a.* 专用的；排外的；高级的；单独的
exempt　*v.* 使免除，豁免　*a.* 免除的，豁免的
fishing gear　渔具；打捞装置
hoover up　扫荡；洗劫
impenetrable　*a.* 不能通过的；顽固的
implement　*v.* 实施，执行；使生效，实现
impunity　*n.* 不受惩罚，无罪；不受损失
jurisdiction　*n.* 管辖权；管辖范围；司法权
lane　*n.* 航道；车道；小路，小巷
lobbyist　*n.* 说客
mandate　*n.*/*v.* 授权；委任；批准；命令；
mandatory　*a.* 强制的；命令的
mangrove　*n.* 红树属树木

mishmash *n.* 混杂物 *v.* 使成为混杂的一堆
moratorium *n.* 暂停，暂禁；[法] 延期偿付权，延缓偿付期
nautical *a.* 海上的，航海的；船舶的
outfit *n.* 全套装备
oversee *v.* 监督
pharmaceutical *a.* 制药的；配药的 *n.* 药物
photosynthesis *n.* 光合作用
proliferate *v.* （使）扩散；（使）增生
provision *n.* 规定，条项，条款
pteropods *n.* 翼足类
quota *n.* （复数 quotas）定量；定额；指标；摊派
sacrosanct *a.* 极其神圣的，不可侵犯的
salient *a.* 显著的，突出的；重要的，主要的
signatory *a.* 签字人的，签约国的 *n.* 签字人，签约
staggering *a.* 难以置信的，令人震惊的
streamline *v.* 把…做成流线型；使简单
subsidize *v.* 以津贴补助
suffocate *v.* （使）窒息而死；阻碍
surge *v.* 蜂拥而来；汹涌；起大浪
surveillance *n.* 盯梢，监督；管制，监视
swamp *n.* 湿地；沼泽（地）
tangle *n.* 纠缠；混乱；争论 *v.* （使）缠结；使陷入，捕获；使纠纷
territorial *a.* 领土的；区域的；陆地的；地球的
transparent *a.* 透明的
trawl *v.* 用拖网捕鱼
treaty *n.* 条约；协议，协商；谈判
undermine *v.* 逐渐削弱；从根基处破坏

Ⅱ. Lead-in & Reading 课文导读

Governing the High Seas

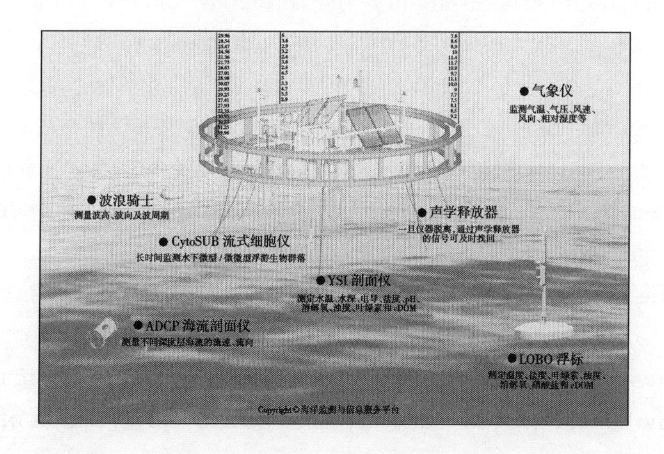

(1) Humans are damaging the high seas[1]. Now the oceans are doing harm back.

(2) Mining is about to begin under the seabed in the high seas—the regions outside the exclusive economic zones (EEZ)[2] administered by coastal and island nations, which stretch 200 nautical miles (370 km) offshore. Nineteen exploratory licenses have been issued. New summer shipping lanes are opening across the Arctic Ocean.

(3) But these developments are minor compared with vaster forces reshaping the Earth, both on land and at sea. It has long been clear that people are damaging the oceans—witness the melting of the Arctic ice in summer, the spread of oxygen-starved dead zones and the death of coral reefs. Now, the consequences of that damage are starting to be felt onshore. Thailand provides a vivid example. In the 1990s it cleared coastal mangrove swamps to set up shrimp farms. Ocean storm surges in 2011, no longer cushioned by the mangroves, rushed in to flood the country's industrial heartland, causing billions of dollars of damage.

(4) The forests are often called the lungs of the Earth, but the description better fits the oceans. They produce half the world's supply of oxygen, mostly through photosynthesis by aquatic algae and other organisms. But according to a forthcoming report by the Intergovernmental Panel on Climate Change (IPCC), concentrations of chlorophyll (which helps make oxygen) have fallen by 9%–12% in 1998–2010 in the North Pacific, Indian and North Atlantic Oceans.

(5) The high seas exemplify the "tragedy of the commons"[3]—the depletion of commonly held property by individual users, who harm their own long-term interests as a result. For decades scientists warned that the European Union's fishing quotas were too high, and for decades fishing lobbyists persuaded politicians to ignore them. Now what everyone knew would happen has happened: three-quarters of the fish stocks in European waters are over-exploited and some are close to collapse.

(6) The salient feature of such a tragedy is that the full cost of damaging the system is not borne by those doing the damage. This is most obvious in fishing, but goes further. Invasive species of many kinds are moved around the world by human activity—and do an estimated $100 billion of damage to

oceans each year. Farmers dump excess fertilizer into rivers, which finds its way to the sea; there cyanobacteria (blue-green algae) feed on the nutrients, proliferate madly and reduce oxygen levels, asphyxiating all sea creatures. In 2008, there were over 400 "dead zones" in the oceans. Polluters pump out carbon dioxide, which dissolves in seawater, producing carbonic acid. That in turn has increased ocean acidity by over a quarter since the start of the Industrial Revolution. In 2012, scientists found pteropods (a kind of sea snail) in the Southern Ocean with partially dissolved shells.

(7) It is sometimes possible to preserve commons by assigning private property rights over them, thus giving users a bigger stake in their long-term health. That is being tried in coastal and island nations' EEZs. But it does not apply on the high seas. Under international law, fishing there is open to all and minerals count as "the common heritage of mankind". Here, a mishmash of international rules and institutions determines the condition of the watery commons.

(8) The high seas are not ungoverned. Almost every country has ratified the UN Convention on the Law of the Sea (UNCLOS)[4], which, in the words of Tommy Koh, president of UNCLOS in the 1980s, is "a constitution for the oceans". It sets rules for everything from military activities and territorial disputes (like those in the South China Sea) to shipping, deep-sea mining and fishing. Although it came into force only in 1994, it embodies centuries-old customary laws, including the freedom of the seas, which says the high seas are open to all. UNCLOS took decades to negotiate and is sacrosanct. Even America, which refuses to sign it, abides by its provisions.

(9) But UNCLOS has significant faults. It is weak on conservation and the environment, since most of it was negotiated in the 1970s when these topics were barely considered. It has no powers to enforce or punish. America's refusal to sign makes the problem worse: although it behaves in accordance with UNCLOS, it is reluctant to push others to do likewise.

(10) Specialized bodies have been set up to oversee a few parts of the treaty, such as the International Seabed Authority, which regulates mining beneath the high seas. But for the most part UNCLOS relies on member countries and existing organizations for monitoring and enforcement. The result is a baffling tangle of overlapping authorities that is described by the

Global Ocean Commission, a new high-level lobby group, as a "coordinated catastrophe".

(11) Individually, some of the institutions work well enough. The International Maritime Organization, which regulates global shipping, keeps a register of merchant and passenger vessels, which must carry identification numbers. The result is a reasonably law-abiding global industry. It is also responsible for one of the rare success stories of recent decades, the standards applying to routine and accidental discharges of pollution from ships. But even it is flawed. The Institute for Advanced Sustainability Studies, a German think-tank, rates it as the least transparent international organization. And it is dominated by insiders: contributions, and therefore influence, are weighted by tonnage.

(12) Other institutions look good on paper but are untested. This is the case with the seabed authority, which has drawn up a global regime for deep-sea mining that is more up-to-date than most national mining codes. For once, therefore, countries have settled the rules before an activity gets under way, rather than trying to catch up when the damage starts, as happened with fishing.

(13) The problem here is political rather than regulatory: how should mining revenues be distributed Deep-sea minerals are supposed to be "the common heritage of mankind". Does that mean everyone is entitled to a part and how to share it out?

(14) The biggest failure, though, is in the regulation of fishing. Overfishing does more damage to the oceans than all other human activities there put together. In theory, high-seas fishing is overseen by an array of regional bodies. Some cover individual species, such as the International Commission for the Conservation of Atlantic Tunas (ICCAT). Others cover fishing in a particular area, such as the northeast Atlantic or the South Pacific Oceans. They decide what sort of fishing gear may be used, set limits on the quantity of fish that can be caught and how many ships are allowed in an area, and so on.

(15) Here, too, there have been successes. Stocks of northeast Arctic cod (*Gadus morhua*) are now the highest of any cod species and the highest they have been since 1945—even though the permitted catch is also at record

levels. This proves it is possible to have healthy stocks and a healthy fishing industry. But it is a bilateral, not an international, achievement: only Norway and Russia capture these fish and they jointly follow scientists' advice about how much to take.

(16) There has also been some progress in controlling the sort of fishing gear that does the most damage. In 1991 the UN banned drift nets longer than 2.5 km. A series of national and regional restrictions in the 2000s placed limits on "bottom trawling" (hoovering up everything on the seabed) —which most people at the time thought unachievable.

(17) But the overall record is disastrous. Two-thirds of fish stocks on the high seas are over-exploited—twice as much as in parts of oceans under national jurisdiction. Illegal and unreported fishing is worth $10 billion - 24 billion a year—about a quarter of the total catch. According to the World Bank, the mismanagement of fisheries costs $50 billion or more a year, meaning that the fishing industry would reap at least that much in efficiency gains if it were properly managed.

(18) Most regional fishery bodies have too little money to combat illegal fishermen. They do not know how many vessels are in their waters because there is no global register of fishing boats. Their rules only bind their members; outsiders can break them with impunity. An expert review of ICCAT concluded that it was "an international disgrace". A survey by the FAO[5] found that over half the countries reporting on surveillance and enforcement on the high seas said they could not control vessels sailing under their flags. Even if they wanted to, then, it is not clear that regional fishery bodies or individual countries could make much difference.

(19) But it is far from clear that many really want to. Almost all are dominated by fishing interests. The exceptions are the organization for Antarctica, where scientific researchers are influential, and the International Whaling Commission, which admitted environmentalists early on. Not by coincidence, these are the two that have taken conservation most seriously.

(20) Reforming specific policies will not be enough. Countries also need to improve the system of governance. There is a basic law of the sea signed by most nations (though not America, to its discredit). But it contains no mechanisms to enforce its provisions. Instead, dozens of bodies have sprung

up to regulate particular activities, such as shipping, fishing and mining, or specific parts of the oceans. The mandates overlap and conflict. And no one looks after the oceans as a whole.

(21) A World Oceans Organization should be set up within the UN. After all, if the UN cannot promote collective self-interest over the individual interests of its members, what is it good for? Such an organization would have the job of streamlining the impenetrable institutional tangle. But it took 30 years to negotiate the law of the sea. A global oceans body would probably take longer—and the oceans need help now.

(22) So in the meantime the law of the sea should be beefed up. It is a fine achievement, without which the oceans would be in an even worse state. But it was negotiated in the 1970s before the rise of environmental concerns, so contains little on biodiversity. And the regional fishing bodies, currently dominated by fishing interests, should be opened up to scientists and charities. As it is, the sharks are in charge of the fish farm.

(23) This would not solve all the problems of the oceans. But institutional reform for the high seas could cut overfishing and, crucially, change attitudes. The high seas are so vast and distant that people behave as though they cannot be protected or do not need protection. Neither is true. Humanity has harmed the high seas, but it can reverse that damage. Unless it does so, there will be trouble brewing beneath the waves.

Notes:

[1] high seas　公海。1982年《联合国海洋法公约》规定公海是不包括在国家的专属经济区、领海或内水或群岛国的群岛水域以内的全部海域。公海供所有国家平等地共同使用。它不是任何国家领土的组成部分，因而不处于任何国家的主权之下，任何国家不得将公海的任何部分据为己有。

[2] exclusive economic zone (EEZ)　专属经济区，又称经济海域，是国际公法中为解决国家或地区之间的因领海争端而提出的一个区域概念。专属经济区是指领海以外并邻接领海的一个区域，从测算领海宽度的基线量起，不应超过200海里（370.4千米），除去离另一个国家更近的点。沿海国对这一区域内的自然资源享有主权权利和其他管辖权，而其他国家享有航行和飞越自由等。该公约签署批准国已达160国。目前仅有美国、朝鲜、伊朗、泰国、利比亚等尚未批准。

[3] tragedy of the commons　公地悲剧。公地作为一项资源或财产有许多拥有者，他们中的每一个都有使用权，但没有权力阻止其他人使用，而每一个人都倾向于过度使用，从而造成资源的枯竭。过度砍伐的森林、过度捕捞的渔业资源及污染严重的河流和空气，

都是公地悲剧的典型例子。之所以叫悲剧，是因为每个当事人都知道资源将由于过度使用而枯竭，但每个人对阻止事态的继续恶化都感到无能为力，而且都抱着“及时捞一把”的心态加剧事态的恶化。

[4] the UN Convention on the Law of the Sea（UNCLOS）　联合国海洋法公约。UNCLOS 指联合国曾召开的三次海洋法会议，以及 1982 年第三次会议所决议的海洋法公约（LOS）。在中文语境中，海洋法公约一般是指 1982 年的决议条文。此公约对内水、领海、临接海域、大陆架、专属经济区、公海等重要概念做了界定。对当前全球各处的领海主权争端、海上天然资源管理、污染处理等具有重要的指导和裁决作用。

[5] FAO（Food and Agriculture Organization of the United Nations）　联合国粮食及农业组织。联合国系统内最早的常设专门机构，其宗旨是提高人民的营养水平和生活标准，改进农产品的生产和分配，改善农村和农民的经济状况，促进世界经济的发展并保证人类免于饥饿。1943 年 5 月根据美国总统 F. D. 罗斯福倡议在美国召开有 44 个国家参加的粮农会议，决定成立粮农组织筹委会，拟订粮农组织章程。1945 年 10 月 16 日粮农组织在加拿大魁北克正式成立，1946 年 12 月 14 日成为联合国专门机构。总部设在意大利罗马。

Ⅲ. Teaching Focus（12）：Proofreading　翻译技巧（12）：校对

一般来说，一个完整的翻译过程需要经过三个阶段：理解、表达和校对。理解就是通读原文，领略原作的内容；表达就是选择恰当的目标语，把已经了解的原作内容表述出来；校对是对理解和表达这两个阶段的内容进行审核，找出其中不确切或不完整的地方，进行修改或完善。翻译的这三个阶段相辅相成，相得益彰，缺一不可。其中，理解是表达的基础，表达是理解的结果，校对则是理解和表达的进一步深化，是对原文内容的进一步核实和对译文语言的进一步推敲。

校对的目的和作用是为翻译质量把关。一部优秀的译作是译者字斟句酌、反复推敲的结果。理想的译作应该像一块透明的玻璃，透过它可以清晰地看到原作，但又注意不到玻璃的存在。但由于原文和译文使用的是两种完全不同语言体系，所承载的文化之间又存在巨大的差异，因此译者无论是在理解还是在表达时都会受到原文的影响或约束，译文质量难免会有这样或那样的缺陷。校对环节正是着眼于这些缺陷，目的就是要消灭那些在翻译过程中出现的错漏或不符合翻译标准的地方，至少应把它们减少到最低。因此我们说，校对是确保译文质量必不可少的、至关重要的环节。翻译的校对工作必须得到应有的重视。

翻译校对需要注意的问题可能会因翻译内容和主题的不同而有一些细微的差别。对于科技翻译（如海洋科技文献翻译）来说，校对阶段应特别注意的问题主要包括以下几点：①检查理解过程中有无错漏，特别是检查对相关专业知识的理解是否准确；②核实译文在人名、地名、日期、数字、方位、度量衡单位、相关性等方面有无错漏，检查专业词汇或术语以及缩略词翻译是否标准或前后一致；③校核译文段落、标点符号、图表代码等是否正确无误；④核实译文是否准确转述原作内容和思想，反思译文语言表达是否规范，是否符合目标语习惯，是否通顺易懂等。其中，最后一点尤其需要用心推敲，也是最费时费力的一项工作。

通常校对最少要进行两遍。第一遍着重核对内容，第二遍着重润饰文字。润饰文字是一个细致活，需要校译者处处审慎、认真核实、通盘处理。在具体操作时，应从以下方面进行核校。

(1) 核校词语搭配是否得当。词语搭配多种多样，有主谓搭配、动宾搭配、修饰语与中心词之间的搭配等。无论哪种搭配方式，都有一定的约定俗成性，有时不得自由组合。而且词语搭配是以目标语为标准，不受原文的约束。

(2) 检查用词是否到注意语义色彩。词语有多种语义色彩，如褒贬色彩、文体色彩等。在翻译时，应全面考虑词义才能做到用词恰当。

(3) 检查用词用语是否会产生歧义。歧义是自然语言中不可避免的现象，但在翻译中应尽量避免。而对于科技文章翻译来说，歧义的表达方式绝对是禁忌，因为这与科技文体严谨、科学、客观、准确的特征背道而驰。

(4) 检查是否有翻译腔。中西文语言有很大差异。汉语以意统形，词约义丰；西文语言重形合，追求形式丰满。用汉族的意合语言译西方的形合语言必然要做结构上的调整，否则就会出现翻译腔。当然，去除翻译腔并不是说凡是翻译都应该采用意译。实际上，对于科技翻译来说，在能确切地表达原作思想内容和不违背译文语言规范的条件下，采用直译手段更可取。因此，翻译工作是经验积累的过程。

(5) 检查译文是否符合逻辑，上下文是否连贯。翻译是逻辑活动，检验译文也应该用逻辑分析的方法进行。如果译文不合情理、自相矛盾、违反常识，那必定包含着译者对原文文理、词汇、语法、修辞意义的错误理解。另外，前后贯通、上下一致，是文章的最基本要求之一。译文的上下文不连贯，所叙述的事实前后不相符，那就隐伏着译者对原文理解的错误。在多人合译篇幅较长的作品时，由于对上下文不甚了然，最容易出现这类错误。因此，核校者需要从整体上把握全文，对上下文要做到心中有数。

(6) 检查语体风格是否协调。一篇文章的语体风格应该在总体上和谐协调。如果在译文的某些句段中突然“跳”出来一个在语气、风格上与整体不协调的字眼，那就得特别注意。

翻译的校对工作又称为校译，以区别于普通意义上的编辑校对工作，这是因为翻译校对者不但需要根据原文，来检查和核对别人或自己的翻译是否有错误，或有哪些不准确以及不通顺的地方，还必须有针对性地对翻译提出改进建议或直接对译文进行修改。这样的校对工作事实上是理解过程和表达过程的重复和升华。因此，从某种意义上来说，如果校对工作和初译工作分别由不同的人来承担的话，那么对校对者的能力要求应该比对初译者的能力要求更高。

做好校译工作要具备多方面的素养。从海洋文献英汉互译工作来说，首先要具有较强的英语和汉语基础，掌握一定数量的专业词汇和语法知识，能够熟练运用两种语言。这是保证校译质量的基本条件。其次，应具备一定的科学常识和较宽的知识面，最好是具有相关专业或学科的基础。但实际情况是，很多翻译工作者都是在进行跨专业或学科的翻译工作。此时，需要他们保持对所译学科和专业的相关知识不断学习和探究的精神。只有了解了更多的专业知识，才能翻译出更专业的译作。最后，校译者还应具有严谨认真的工作作风。“文章不厌百回改”，主体虽然说的是作者，但这句话同样适用于从事翻译校对工作的校译者。校译的过程是一个反复推敲、不断切磋的过程，需要译者具有一丝不苟、不畏困难和锲而不舍的精神。

Questions for Discussion:

1. What are the purposes of proofreading?

2. Are there any other elements that are important in proofreading, yet not mentioned in this unit?

Ⅳ. Exercise 1: Sentences in Focus　练习 1：单句翻译

1. Translate the following sentences into Chinese

(1) Ocean storm surges in 2011, no longer cushioned by the mangroves, rushed in to flood the country's industrial heartland, causing billions of dollars of damage.

(2) The high seas exemplify the "tragedy of the commons"—the depletion of commonly held property by individual users, who harm their own long-term interests as a result.

(3) The salient feature of such a tragedy is that the full cost of damaging

the system is not borne by those doing the damage. This is most obvious in fishing, but goes further.

(4) Farmers dump excess fertilizer into rivers, which finds its way to the sea; there cyanobacteria (blue-green algae) feed on the nutrients, proliferate madly and reduce oxygen levels, asphyxiating all sea creatures.

(5) It is sometimes possible to preserve commons by assigning private property rights over them, thus giving users a bigger stake in their long-term health.

(6) Although it came into force only in 1994, it embodies centuries-old customary laws, including the freedom of the seas, which says the high seas are open to all.

(7) The result is a baffling tangle of overlapping authorities that is described by the Global Ocean Commission, a new high-level lobby group, as a "coordinated catastrophe".

(8) And it is dominated by insiders: contributions, and therefore influence, are weighted by tonnage.

(9) Stocks of north-east Arctic cod are now the highest of any cod species and the highest they have been since 1945—even though the permitted catch is also at record levels.

(10) A series of national and regional restrictions in the 2000s placed limits on "bottom trawling" (hoovering up everything on the seabed) —which most people at the time thought unachievable.

(11) According to the World Bank, the mismanagement of fisheries costs $50 billion or more a year, meaning that the fishing industry would reap at least that much in efficiency gains if it were properly managed.

(12) Their rules only bind their members; outsiders can break them with impunity. An expert review of ICCAT concluded that it was "an international disgrace".

(13) Instead, dozens of bodies have sprung up to regulate particular activities, such as shipping, fishing and mining, or specific parts of the oceans.

(14) After all, if the UN cannot promote collective self-interest over the individual interests of its members, what is it good for?

(15) And the regional fishing bodies, currently dominated by fishing

interests, should be opened up to scientists and charities. As it is, the sharks are in charge of the fish farm.

2. Translate the following sentences into English

(1) 海洋生物的基因资源保证了制药业的财源，相关的专利数量一直以每年12%的幅度上升。

(2) 当前，我国公共池塘资源的开发中一定程度地存在“逆向选择”问题，它最终会导致公地悲剧的发生。

(3) 为了维持地球生命支持系统，提供粮食和生计、对人类的福利做出贡献，并确保经济的可持续发展，需要有健康的海洋和陆地生态系统及它们所提供的服务。

(4) 环保组织还敦促各国政府打击渔业管理部门内部的贿赂和腐败行为。这些行为带动了非法捕鱼，如让一些无执照的渔船免受严格监督。

(5) 在东海，捕虾桁拖网作业被强制暂停，因为桁拖网网具对鱼类个体差的选择性，不利于渔业资源保护，也给渔业管理带来了负面影响。

(6) 如今随着全球变暖和能源安全问题被提上欧美及亚洲政府议事日程，核能、生物燃料、节能技术及其他油气替代品开始受到越来越多的关注。

(7) 在他们管辖的航道内，任何人不得弃置沉船，不得放置碍航渔具，不得种植水生植物。

(8) 金枪鱼是洄游于公海和沿海国家200海里经济专属水域内的大洋性鱼类。

(9) 好的抗氧剂应该能在一定程度上把空气和皮肤隔离开，但决不会窒息皮肤。

(10) 研究人员担心，由于大气中二氧化碳水平的升高，海水不断变酸，许多无脊椎动物会因此而消失。

(11) 我们在预测和避免危机方面的集体失误，给我们的较长远发展带来了重要教训。

(12) 我们要加强经济技术合作，推动区域贸易和投资自由化，缩小成员发展差距，实现共同繁荣。

(13) 实施补助金计划是为了减轻渔民因难以预测的恶劣天气而遭受的损失。

(14) 似乎只有当环境恶化或资源枯竭导致了经济衰退时，我们才会注意到问题的严重性。

(15) 所有功能正常的政治体系可能是相似的，但每个功能失调的政府各有各的失调方式。

Ⅴ. Exercise 2: Passages in Focus 练习 2: 短文翻译

1. Translate the following passages into Chinese

Countries could do more to stop vessels suspected of illegal fishing from docking in their harbors—but they don't. The FAO's attempt to set up a voluntary register of high-seas fishing boats has been becalmed for years. The UN has a fish-stocks agreement that imposes stricter demands than regional fishery bodies. It requires signatories to impose tough sanctions on ships that break the rules. But only 80 countries have ratified it, compared with the 165 parties to UNCLOS. One study found that 28 nations, which together account for 40% of the world's catch, are failing to meet most of the requirements of an FAO code of conduct which they have signed up to.

It is not merely that particular institutions are weak. The whole system itself is dysfunctional. For example, there are organizations for fishing, mining and shipping, but none for the oceans as a whole. Regional seas organizations, whose main responsibility is to cut pollution, generally do not cover the same areas as regional fishery bodies, and the two rarely work well together. Dozens of organizations play some role in the oceans (including 16 in the UN alone) but the outfit that is supposed to co-ordinate them, called UN-Oceans, is an ad-hoc body without oversight authority. There are no proper arrangements for monitoring, assessing or reporting on how the various organizations are doing—and no one to tell them if they are failing.

According to David Miliband, a former British foreign secretary who is now co-chairman of the Global Ocean Commission, the current mess is a "terrible betrayal" of current and future generations. "We need a new approach to the economics and governance of the high seas," he says.

That could take different forms. Environmentalists want a moratorium on overfished stocks, which on the high seas would mean most of them. They also want regional bodies to demand impact assessments before issuing fishing licenses. The UN Development Program says rich countries should switch some of the staggering $35 billion a year they spend subsidizing fishing on the high seas (through things like cheap fuel and vessel-buy-back programs) to creating marine reserves—protected areas like national parks.

Others focus on institutional reform. The European Union and 77 developing countries want an "implementing agreement" to strengthen the environmental and conservation provisions of UNCLOS. They had hoped to start what will doubtless be lengthy negotiations at a UN conference in Rio de Janeiro in 2012. But opposition from Russia and America forced a postponement.

Still others say that efforts should be concentrated on improving the regional bodies, by giving them more money, greater enforcement powers and mandates that include the overall health of their bits of the ocean. The German Advisory Council on Global Change, a think-tank set up by the government, argues for an entirely new UN body, a World Oceans Organization, which it hopes would increase awareness of ocean mismanagement among governments, and simplify and streamline the current organizational tangle.

2. Translate the following passages into English

1968 年美国生态学家加勒特哈丁发表了一篇名为"公地悲剧"的文章。他指出，当资源共有时，耗尽它是符合个人利益的，因此人们倾向于通过过度开发的方式破坏自身长远的集体利益，而不是保护这一资源。这种悲剧正在上演，给覆盖地球约一半面积的一种资源造成了严重危害。

公海——位于沿海国家 200 海里专属经济区之外的海域——就是一种公地。所有人都有权在那儿捕鱼；各国已宣布公海海床上的矿物为"人类共同的遗产"。公海对每一个人来说都经济意义重大，而且随着鱼肉成为一种比牛肉更重要的蛋白质来源，其重要性日益增显。利用海洋生物 DNA 发明的专利数量迅速上升；并且一项研究表明，海洋生物含有抗癌药有用物质的可能性多出陆生生物一百倍。

然而，公海的状况正在恶化。现在，北极冰川夏天会融化；死区正在蔓延。三分之二的鱼类资源被过度开发，其严重程度甚至超过国家控制下的海洋部分。怪异的事情正在微生物层面发生。海洋能产出地球氧气供应量的一半，主要是多亏了水生藻类的叶绿素。而现在，叶绿素含量正在下降。情形还不至于使生命窒息，但它可能进一步危害气候，因为氧气的减少就意味着二氧化碳的增多。

为了避免公地悲剧，我们需要一些规章制度来平衡个人的短期利益与所有人的长远利益。这就是为什么需要对那些功能失调的公海管理政策和机构进行彻底改革，以减少净损失。

第一个改革目标应该是渔业补贴。渔民在一个国家的自我形象中常居重要地位，他们成功说服了政府用其他人的钱来补贴一个损失数十亿美元且造成巨大环境破坏的产业。富裕国家以廉价燃油、保险等方式，每年给这些耗损公海资源的人们 350 亿美元的补助，总数额超过了其捕获价值的三分之一。这种做法应该停止。

第二，应该对全球渔船进行注册。一项国际规定要求客船和货船必须持有独有的身份号码，而渔船长期以来一直免于这项要求。去年 12 月各海洋国家解除了该项豁免，迈出了良好的第一步。但各国仍自行决定是否要求悬挂本国国旗的渔船参与身份认证计划。政府应强制执行这项计划，创建一个全球性的船只记录数据，以协助打击非法公海捕鱼。

第三，应建立更多的海洋自然保护区。地球上八分之一的陆地享有法律保护措施（如国家公园），而仅有不到百分之一的公海享有这样的保护。在过去几年里，各国已开始在他们自己的经济区内建立海洋保护区。公海捕鱼管理机构应复制这种理念，为鱼类种群和环境提供一些恢复的空间。

Appendix 1 Glossary

附录 1 词汇表

A a

abate *v.* 减少；[法] 取消法令 *n.* 减轻；折扣 U11
abet *v.* 煽动；教唆；怂恿 U11
accommodate *v.* 容纳 U5
accountable *a.* 负有责任 U6
accretion *n.* 添加，冲积层 U3
acid gas 酸性气（体） U9
acid rain 酸雨 U9
acidify *v.* 酸化 U4
adhere to 黏附 U4
ad-hoc *n.* 点对点；特设；专门 U12
adjacent *a.* 邻近的，毗邻的 U8
admiral *n.* 海军上将；舰队司令；将军 U10
advent *n.* 到来，出现 U7
adversity *n.* 逆境；不幸；灾难 U11
advisable *a.* 明智的；可取的；适当的 U6
afflict *v.* 使受痛苦；折磨 U5
algae *n.* 藻类 U4
alien *a.* 外国的；相异的 *v.* 疏远；离间 U11
alleviate *v.* 减轻，缓和 U11
allocation *n.* 配给，分配；分配额 U6
allure *n.* 诱惑 U7
anchoveta *n.* 南美鳀 U2
anchovy *n.* 鳀鱼 U2
anoxic *a.* 缺氧的 U4
Antigua *n.* 安提瓜岛 U7

apparatus *n.* 仪器；器械；装置 U4
aquaculture *n.* 水产养殖；水产业 U6
aquatic *a.* 水生的；水产的 *n.* 水生动植物 U4
arc *n.* 弧线（复数为 arcs）；弧形物 U10
archipelagic *a.* 群岛的 U8
armada *n.* 舰队 U8
array *n.* 排列，大批 U3
artery *n.* 干线，要道；[解剖] 动脉 U10
asphyxiate *v.* （使）窒息 U12
atmosphere *n.* 大气，大气层，大气圈 U1
attendant *a.* 伴随的 U3
auditor *n.* 审计员；查账员 U6
autonomy *n.* 自主权；自治，自治权 U6
auxiliary *a.* 辅助的；备用的 U5
avert *v.* 转移；防止，避免 U12

B b

Bab el Mandeb 曼德海峡 U8
backhaul *n.* 返程；回程运费；回运 U10
baffling *a.* 令人困惑的，难对付的，难解的 U12
Bahamas *n.* 巴哈马群岛 U7
ballast *n.* 镇流器；压舱物，压载物 U4
Baltic Sea （欧洲）波罗的海 U6
bank *n.* 浅滩 U2
Barbados *n.* 巴巴多斯（拉丁美洲国家） U7
Barbuda *n.* 巴布达岛 U7
barge *n.* 驳船；游艇；游览汽车 U10
barrage *n.* 拦河坝 U9
basalt *n.* 玄武岩 U1
baseline *n.* 基准线，零位线 U6
bauxite *n.* 矾土 U7
bauxite *n.* 铝土矿，铝矾土 U10
be geared to/towards 适应；适合 U11
becalm *v.* 使（船等）停住 U12

beef up 加强，补充；使更大（更好等） U12
beneficiary *n.* 受益者 U8
benthic *a.* 海底的，底栖的 U3
benthos *n.* 海底生物 U4
bilateral *a.* 双边的；双方的；两侧的 U12
biodegrade *v.*（进行）生物递降分解；自然降解 U4
biomass *n.*（单位面积或体积内）生物量 U2
biopolymer *n.* 生物高聚物 U11
bonanza *n.*（突然的）财源；富矿脉 U12
Bornholm *n.*（丹麦）博恩霍尔姆岛 U6
botanical garden 植物园 U7
boulders *n.* 卵石，圆石；巨砾，冰砾；漂砾 U1
brackish water 半咸水，微咸水 U11
break-bulk cargo 杂货；杂货运输 U10
breed *v.* 交配；产仔；繁殖 *n.* 种类；血统 U11
brew *v.* 即将发生；酿造；泡，煮；策划 U12
bucket system 斗式系统 U4
bulk cargo 散货；散货运输 U10
bulldoze *v.* 用推土机清除 U2
bureaucracy *n.* 官僚主义；官僚机构 U6
bycatch *n.* 兼捕 U2

C c

Canadian Coast Guard (CCG) 加拿大海岸警卫队 U5
Cape Cod 科德角 U2
capsize *v.* 使（船或车）翻；倾覆 U5
captivity *n.* 俘虏；圈养 U2
capture fishery 捕捞渔业 U11
carbon dioxide 二氧化碳 U1
cargo *v.*（船或飞机装载的）货物 U5
carnival *n.* 狂欢节 U7
casualty *n.* 伤亡（人数）；事故 U5
catalyst *n.* 催化剂，刺激因素 U7
catastrophic *a.* 灾难的；惨重的 U5

catch *n.* 捕获量 U4
cater to 迎合，为…服务 U7
catfish *n.* 鲶鱼 U11
chlorophyll *n.* 叶绿素 U2
citrus *n.* 柑橘 U7
classify *v.* 分类，归类；把…列为密件 U6
clog *v.* 阻塞，阻碍；闭合 U4
coalesce *v.* 使联合，使合并 U1
Coast Guard 海岸警卫队 U5
Coast Guard Auxiliary (CGA) 辅助海岸警卫队 U5
coastline *n.* 海岸线 U1
cobalt *n.* 钴 U4
cobalt crust 富钴结壳 U3
cod *n.* 鳕鱼 U2
cohort *n.* 世代；年龄组 U2
collision *n.* 碰撞；冲突 U5
comets *n.* 彗星 U1
commercial-scale 商业规模的，商业性的 U9
commoditize *v.* 商品化 U7
conceivable *a.* 可想到的，可相信的 U5
configuration *n.* 配置，结构，外形 U3
constancy *n.* 稳定性；持续性；持久性 U6
containerization *n.* 货柜运输；集装箱运输 U10
contemporarily *adv.* 同时期地，同时代地 U10
contiguous *a.* 邻近的，共同的 U6
continent *n.* 大陆；洲；(the Continent) 欧洲大陆 U1
continental crust 陆壳 U1
continental rocks 大陆岩石 U1
contingency *n.* 应急；偶发事件 U5
converge *v.* 聚集；会于一点；接轨 U10
coordinator *n.* 协调员 U5
copepods *n.* 桡脚类动物 U4
coral polyps 珊瑚虫 U4
coral reef 珊瑚礁 U12

core *n.* 地核，核心 U1
cornerstone *n.* 奠基石；基础 U6
corrosion *n.* 侵蚀，腐蚀 U4
cottage industry 家庭手工业 U8
crack down on/upon 打击；制裁；镇压 U12
cruise *n.* 巡游，巡航，乘船游览 U7
crust *n.* 地壳；外壳 U1
crustacean *n.* 甲壳纲动物 U2
crustal rocks 地壳岩石 U1
crystalline *a.* 结晶状的，透明的 U3
culture *v.* 培植，培养 *n.* 养殖 U11
curative *a.* 有疗效的，治病的 U7
cushion *n.* 垫子；起保护（或缓冲）作用的事物 *v.* 起缓冲作用 U12
cyanobacteria *n.* 蓝细菌；蓝藻 U12
cyprinid *n.* 鲤科鱼类 *a.* 鲤科的 U11

D d

dairy *a.* 乳制品；奶品 *n.* 乳品业；牛奶场 U4
deadweight *n.* 重负，重担；吨位 U10
deep depression 深凹区，深坳陷 U1
degas *v.* 排气，排除煤气 U1
delegate *v.* 委派代表；授权给 U5
Department of National Defence （加拿大）国防部 U5
dependant *a.* 随…而变的 U5
depletion *n.* 耗竭，消耗，用尽 U8
depletive *a.* 使干涸的；使折损的 U4
deposit *n.* 矿床，沉积物 *v.* 沉淀 U3
designate *v.* 指出；指派 U5
designation *n.* 指定 U8
deteriorate *v.* （使）恶化 U12
detour *n.* 绕道，迂回 U8
detrimental *a.* 有害的，不利的 U8
deviation *n.* [航] 偏航；背离；离经叛道的行为 U10
dhow *n.* 单桅三角帆船 U8

diagram *n.* 图表 U6
diffuse *a.* 四散的，散开的 U4
digestive tract 消化道 U4
dilemma *n.* 窘境，困境 U6
disastrous *a.* 灾难性的；极坏的；悲惨的 U12
discharge *n.*/*v.* 流出；排放 U4
discount rate 贴现率，折现率 U9
disintegrate *v.* （使某物）碎裂，崩裂，瓦解 U4
disjoint *v.* （使）脱节，（使）解体 U6
dissolution *n.* 溶解，融化 U4
dissolve *v.* （使）溶解；液化；消失；终止 U12
distillery *n.* 酿酒厂 U7
dominant *a.* 占优势的；统治的 U6
dominion *n.* 主权，统治权 U8
draft *n.* 草稿；拟量 U10
dredge *v.* 疏浚；挖掘；使显露；采捞 U10
dredging *n.* 清淤 U2
driftnetting *n.* 流刺网捕鱼 U2
drum *n.* 桶；鼓 U10
dugong *n.* 儒艮属；儒艮，海牛 U4
dysfunctional *a.* 功能失调的；不良的 U12

E e

ebb *n.* 退潮，落潮 *v.* （潮水）退落 U9
ecosystem *n.* 生态系统 U4
efficiency gains 效率利得；效率收入 U12
effluent *a.* 流出的 *n.* 污水；流出的水流 U11
electrical grid 输电网络；电气网格 U9
elevation *n.* 高地，隆起 U3
enactment *n.* 制定，颁布；法律，法规；扮演 U11
encapsulate *v.* 概述 U7
enclave *n.* 被包围的领土（区域）；飞地（指在甲国境内的隶属乙国的一块领土） U10
encompass *v.* 围绕，包围 U6

endemic *a.* 地方性的 U3
endowment *n.* 赠与，天赋 U8
entangle *v.* 使纠缠，缠住；使卷入 U4
entrench *v.* 用壕沟围绕或保护…；牢固地确立 U11
entrepreneur *n.* 企业家 U7
equitable *a.* 合理；公正的 U6
equity *n.* 公平；公道 U11
erosion *n.* 侵蚀，腐蚀 U3
errand *n.* 差事；使命 U5
escort *v.* 护送，护航 U8
estuary *n.* 港湾；（江河入海口）河口 U4
euphoria *n.* 欢快，兴高采烈 U3
eutrophication *n.* 富营养化；超营养作用 U4
evacuation *n.* 撤离；疏散 U5
exclusive *a.* 专用的；排外的；高级的 U12
exempt *v.* 使免除，豁免 *a.* 免除的，豁免的 U12
expeditious *a.* 迅速的，敏捷的 U8
exponentially *adv.* 以指数方式 U4
extraterritorial *a.* 疆界之外的 U3
extrude *v.* 挤出，压出 U3
eye *v.* 注意，看 U8

F f

facilitate *v.* 助长；促成 U4
fauna *n.* 动物群 U7
federal *a.* 联邦（制）的，同盟的 U6
ferromanganese *n.* 锰铁 U3
ferry *n.* 渡船；渡口；摆渡 *v.* 航海；渡运；摆渡 U10
finfish *n.* 长须鲸 U11
fired power plant 火（力发）电厂 U9
fishing gear 渔具；打捞装置 U12
fishmeal *n.* 鱼粉 U11
flank *n.* 侧面，侧翼 U3
flexible *a.* 灵活的；易弯曲的 U5

flora *n.* 植物群 U7
flotilla *n.* 小型船队 U8
flotsam *n.* 废料；漂流残骸 U4
flow velocity 流速 U9
fluvial *a.* 河的，河流的，河中的 U10
flux *n.* 流量；流出；熔解 U4
formulation *n.* 配方；构想，规划；公式化 U11
fossil fuel 化石燃料，矿物燃料 U9
fracture zone 断裂带 U3
fragment *v.*（使）碎裂 *n.* 碎片；片段 U11
framework *n.* 框架，构架；（体系的）结构 U6

G g

gas hydrate 气体水合物 U3
gaseous state 气（体状）态 U1
gear *n.* 齿轮 *v.* 接上；使适应；装上齿轮 U11
gear to 使适合于 U7
geographical *a.* 地理学的，地理的 U6
global warming 全球（气候）变暖 U9
go into effect 生效 U6
golden apple snail 福寿螺 U11
"Goldilocks" planet 适合（人类）居住的行星 U1
granite *n.* 花岗岩 U1
gravel *n.* 碎石；沙砾 U3
gravitational attraction 地球（地心、万有）引力 U1
Great Lakes （美国）五大湖区 U6
greenhouse *n.* 温室，玻璃暖房 U1
Gross Domestic Product (GDP) 国内生产总值 U7
grounding *n.* 搁浅 U5

H h

habitat *n.* 栖息地，产地 U1
haddock *n.* 黑线鳕鱼 U2
handlining *n.* 手钓 U2

harness *n.* 利用，治理，控制 U9
hatchery *n.* （尤指鱼的）孵化场 U11
hazard *n.* 危险；冒险的事 U5
head of water 水头 U9
herring *n.* 鲱鱼 U2
hibernation *n.* 冬眠；过冬 U2
high（low）tide 满（低）潮，高（低）潮（期） U9
hinder *v.* 阻碍，妨碍 U5
hinterland *n.* 腹地；内地；穷乡僻壤 U10
hoover up 扫荡；洗劫 U12
hub *n.* 轮轴；枢纽；插孔；[计] 集线器 U10
hurricane *n.* 飓风 U7
husband *v.* 节约地使用或管理 U8
hydraulic *a.* 水力的，水压的；用水发动的 U9
hydraulic pump 液压泵 U4
hydrocarbon reserve 油气储量 U8
hydrographic *a.* 与水文地理有关的 U10
hydrolysate *n.* 水解产物；酶解物 U4
hydropower *n.* 水力发电 U9
hydrosphere *n.* 水圈，水界，水气 U1
hydrostatic head 静水头，静压头，静水压头 U9
hydrothermal *adj.* 热液的 U3
hydrothermal vents 热液喷口 U4
hydroxide *n.* 氢氧化物 U3
hypoxic *a.* 含氧量低的 U4

I i

ice cap 冰帽，冰冠 U1
impenetrable *a.* 不能通过的；顽固的 U12
implement *v.* 实施，执行；使生效 U12
impoundment *n.* 收集；蓄水 U9
impunity *n.* 不受惩罚，无罪；不受损失 U12
in conjunction with 与…协力 U5
incentive *n.* 动机；诱因 *a.* 刺激的；鼓励的 U11

incentive *n.* 刺激，诱因，激励 U7
Incident Command Centre (ICC) 事故指挥中心 U5
Incident Command System (ICS) 事故指挥系统 U5
incontrovertible *a.* 无可辩驳的，不容置疑的 U11
indigenous *a.* 土生土长的；生来的；本地的 U6
infectious *a.* 有传染性的；易传染的 U5
infrastructure *n.* 基础设施 U7
infrequent *a.* 稀少的；罕见的； U5
infringement *n.* 侵犯，违反 U8
institutionalize *v.* 使…制度化 U11
insular *n.* 海岛的，孤立的，与世隔绝的 U7
integral *n.* 构成整体所必需的 U2
integration *n.* 整合；一体化；混合；集成 U10
intensification *n.* 强化；激烈化；加厚 U10
interagency *n.* 跨部门的 U5
interface *n.* 交接；接口 *v.* 接合；交谈 U5
intermediate *a.* 中间的 *v.* 调解；干涉 *n.* 中间人 U10
intertidal zone 潮间带 U6
intruder *v.* 侵入者 U8
inventory *n.* 目录，清单 U6
itinerary *n.* 旅行日程 U7

J j

jeopardize *v.* 危及；损害 U4
Joint Rescue Co-ordination Center (JRCC) 联合救援协调中心 U5
jurisdiction *n.* 管辖权，司法权，管辖范围 U8
juvenile *a.* 幼年的 U2

L l

lagoon *n.* 环礁湖，潟湖；咸水湖 U7
larvae *n.* 鱼崽 U4
league *n.* 里格（长度单位） U8
legislation *n.* 立法，制定法律 U6
liaison *n.* 联络，联络人 U5

likewise　*adv.* 同样地；也，而且　U6
Lima　*n.* 利马　U8
limestone cave　溶洞　U7
lithosphere　*n.* 岩石圈，地壳　U1
littoral　*a.* 沿海的，滨海的　U8
loadline　*n.* 载重线　U10
lobbyist　*n.* 说客　U12
lodge　v. 暂住，寄存　U7
Lofoten　*n.* 罗弗敦群岛　U2
longlining　*n.* 延绳钓　U2
loop　*n.* 回路；圈，环　*v.*（使）成环，（使）成圈　U10
low-head hydroelectricity　低水头水力发电　U9
lucrative　*a.* 赚钱的，有利可图的　U8
lunar cycle　月（亮）运（行）周期　U9

M m

magma　*n.* 岩浆　U3
magnetic anomaly　磁异常　U1
Major Maritime Disaster Contingency Plan（MMDCP）　重大海难应急预案　U5
malaria　*n.* 疟疾，瘴气　U7
mandate　*n.* / *v.* 授权；委任；批准；命令　U12
maneuverability　*n.* 可操作性，机动性；可控性　U10
manganese　*n.* 锰　U3
mangrove　*n.* 红树属树木　U12
mantle　*n.* 地幔　U1
marine biology　海洋生物学　U1
marine ecosystem　海洋生态系统　U6
marine management　海洋管理　U6
marine protected areas（MPAs）　海洋保护区　U6
market access　市场准入　U11
Mars　*n.* 火星　U1
marsh　*n.* 沼泽；湿地　U9
marshalling　*n.* 召集，集结　U5
mass rescue operation（MRO）　大规模救援行动　U5

massification *n.* 大众化；群众化 U10
mechanical power 机械功率，机械动力 U9
Mediterranean *n.* 地中海 U8
megawatt（MW） *n.* 百万瓦特，兆瓦（特） U9
menhaden *n.* 鲱鱼 U2
mercury *n.* 水银，汞 U4
Mesoamerican *n.* 中美洲 U6
metabolism *n.* 新陈代谢 U2
meteorite *n.* 陨星；陨石；陨铁；流星 U1
methane *n.* 甲烷，沼气 U1
microstate *n.* 微型国家，超小国家 U7
migrant vessel 移民船只 U5
mishmash *n.* 混杂物 *v.* 使成为混杂的一堆 U12
mitigate *v.* （使）缓和；减轻；平息 U11
mitigation *n.* 减缓；减轻；平静 U4
mollusc（mollusk） *n.*〈动〉软体动物 U11
mollusk *n.* 软体动物 U2
molten rock 熔融岩石 U1
Montevideo 蒙得维的亚 U8
moratorium *n.* 暂停，暂禁；[法] 延期偿付权 U12
mortality *n.* 死亡数，死亡率 U2
mudflat *n.* 泥滩 U9
mullet *n.* 鲻鱼 U2
multi-agency *n.* 多机构 U5
multi-casualty incident（MCI） 重大伤亡事故 U5
mutation *n.* 突变，变异 U4
mutatis mutandis 加上必要的变更 U8
MWe (abbr＝megawatts of electricity/meters of water equivalent)
兆瓦电力/水当量米数 U9

N n

National Marine Sanctuaries 国家海洋保护区 U6
nautical *a.* 航海的；船舶的；海员的 U12
navigation *n.* 航行，航海；导航 U9

neo-colonization *n.* 新殖民 U7
niche *n.* 合适的位置；有利可图的缺口，商机 U11
nickel *n.* 镍 U1
nitrogen *n.* 氮 U4
NOAA 〈美〉国家海洋和大气局 U6
nodule *n.* 结瘤，结核 U3
non-aligned *a.* 不结盟的；中立的 U5
NOS（The National Ocean Service） 美国国家海洋局 U6
nuanced *a.* 有细微差别的；微妙的 U11
nuclear plant 核电站，核能发电厂 U9
nucleus *n.* 核，原子核 U3

O o

oblique *a.* 倾斜的，不光明正大的 U7
observation *n.* 观察；观察力 U6
ocean crust 洋壳 U1
ocean current 洋流，海流 U9
ocean thermal energy conversion（OTEC） 海洋热能转换系统 U9
offshoot *n.* 分枝；支流；衍生物 U11
offshore wind 离岸风，海上风 U9
On-Scene Commander（OSC） 海难现场指挥员 U5
onshore *a.* 陆上的，朝着岸上的 U3
optimum *a.* 最佳的 U8
ore *n.* 矿砂，矿石 U3
organism *n.* 有机体；生物体；有机组织 U1
ornamental *a.* 观赏性的；装饰性的 U2
oscillation *n.* 振动；波动；动摇；〈物〉振荡 U4
outfit *n.* 全套装备 U12
overflight *n.* 飞越上空 U8
oversee *v.* 监督 U12
overtax *v.* 使负担过重，使过度疲劳 U5
oyster *n.* 牡蛎 U2
ozone *n.* 臭氧，新鲜空气 U7

P p

panacea *n.* 灵丹妙药，万能药 U7
pangaea *n.* 泛大陆，盘古大陆 U1
pangasiid catfish 华鱼芒 U11
papyrus *n.* 纸莎草 U8
participate *v.* 参加某事；分享某事 U6
pathogen *n.* 病菌，病原体 U11
patrol *v.* 巡逻，巡航 U8
pendulum *n.* 钟摆；摇摆不定的事态（或局面） U10
peril *n.* 危险；冒险 *v.* 置…于危险中；危及 U11
peripheral *a.* 边缘的，外围的 U7
periphery *n.* 边缘地带；外围边界 U4
permanence *n.* 永久，持久 U6
pertinent *a.* 相关的 U6
pharmaceutical *a.* 制药的；配药的 *n.* 药物 U12
phosphorus *n.* 磷 U4
photic zone 透光区 U2
photodegrade *v.*（使）光致分解；（使）光降解 U4
photosynthesis *n.* 光合作用 U12
physiography *n.* 地文学，地相学 U10
phytoplankton *n.* 浮游藻类，浮游植物 U2
piracy *n.* 海上抢劫 U5
planet *n.* 行星 U1
planetary bodies 行星体 U1
planetesimal *n.* 小行星体，星子 *a.* 星子的，星子组成的 U1
plankton *n.* 浮游生物 U4
plate tectonics 板块构造（论） U1
platinum *n.* 铂 U3
Pluto *n.* 冥王星 U1
ply *v.* 定期来往于 U8
pollock *n.* 鳕鱼类 U2
polycyclic aromatic hydrocarbons（PAHs） 多环芳烃 U4
polymerase *n.* 聚合酶 U11

polymers *n.* 高分子化合物；多聚物；聚合物 U4
polymetallic *a.* 多金属的 U3
polymetallic nodules 多金属结核 U4
pore waters 孔隙水 U3
port call 沿途到港停靠 U10
powerhouse *n.* 发电站，发电所 U9
precipitate *n.* 沉淀物 *v.* 沉淀 U3
precipitation *n.* 沉淀；（雨等）降落；降雨量 U4
predominance *n.* 优势；主导或支配的地位 U11
preference *n.* 偏爱；优先权 U6
prejudicial *a.* 有害于，不利于 U8
privileged *a.* 享有特权的；特许的 U6
projectile *n.* 投射物 U8
proliferate *v.* （使）扩散；（使）激增 U12
prominent *a.* 著名的；突出的 U6
prospect *n.* 勘探 *v.* 前景，预期 U3
protein *n.* 蛋白质 U2
protocol *n.* 协议 U6
provision *n.* 规定，条项，条款 U12
proximity *n.* 接近，邻近 U8
pteropods *n.* 翼足类 U12
pumped storage 抽水蓄能 U9
purse seine *n.* 围网 U2

Q q

quarter *n.* 一年中的某一季 U6
quota *n.* （复数 quotas）定额；指标；摊派 U12

R r

radioactive elements 放射性元素 U1
ratify *v.* 批准，许可 U6
receipt *n.* 收入，收据，收到 U7
reciprocity *n.* 互惠；相互作用；互给 U10
recruitment *n.* 补充量 U2

refinery *n.* 精炼设备；提炼厂；冶炼厂 U10
reggae *n.* 雷鬼音乐 U7
regime *n.* 管理体系，制度，政权 U8
renewable energy 可再生能源 U9
renewable source 可再生能源 U9
residue *n.* 残余；残渣；残留物 U4
resilience *n.* 快速恢复的能力 U5
resurgence *n.* 复活，再现 U3
retrieval *a.* 回收的；提取的 U4
revenue *n.* 税收，收益 U7
revitalization *n.* 复苏，振兴 U7
riparian *a.* 沿岸的 U8
robust *n.* 鲁棒（性），稳定（性） U9
rotating disk 旋转圆盘 U1
runoff *n.* 径流 U2

S s

sacrosanct *a.* 极其神圣的，不可侵犯的 U12
sailship *n.* 帆船 U10
salient *a.* 显著的，突出的；重要的 U12
salmon *n.* 鲑鱼 U2
sanatorium *n.* 疗养院 U7
sanction *n.* 约束；制裁；批准 *v.* 批准；鼓励 U6
scallop *n.* 扇贝 U2
Scandinavia *n.* （欧洲）斯堪的纳维亚（半岛） U6
scatter *v.* 分散；撒开 U6
scuba diving 轻便潜水 U7
sea floor 海底；海床 U1
seaboard *n.* 沿海地带；海岸 *a.* 海岸的 U9
seaborne trade 海运贸易 U10
seamanship *n.* 航海技术；船舶驾驶术 U5
seamount *n.* 海山；海峰 U6
Search and Rescue (SAR) 搜救 U5
seasonality *n.* 季节性 U10

sector *n.* 部门；领域 U6
sediment *n.* 沉淀物 U3
sediment *n.* 沉淀物；沉渣 U4
seep *v.* 渗漏 U3
seining *n.* 围网捕捞 U2
semi-permeable *a.* 半渗透的 U9
sensuality *n.* 好色，喜爱感官享受 U7
serenity *n.* 宁静，平和 U7
sewage *n.* 排污，下水道 U7
shellfish *n.* 贝类 U2
shoreline *n.* 海岸线 U5
shoreline *n.* 海岸线，海滨线 U9
shy (away) from 厌恶；害怕；回避 U11
signatory *a./n.* 签字人（的），签约国（的） U3
signature dish 招牌菜，拿手菜 U7
silicate *n.* 硅酸盐 U3
sluice *n.* 水闸，有闸人工水道 U9
smuggle *v.* 私运；走私 U5
smuggler *n.* 走私者，走私船 U8
snorkeling *n.* 浮潜 U7
solar power 太阳能 U9
solar radiation 太阳辐射 U1
solution *n.* 溶液 U3
sovereign *n.* 主权国 U8
span *v.* 持续，包括 U6
spectacular *a.* 惊人的；场面壮观的；引人注意的 U11
spill *n.* 泄露；溢出 U5
sports gear 运动器材 U7
spur *n.* 马刺；激励因素 *v.* 鞭策；激励；推动 U11
squid *n.* 鱿鱼 U2
St. Lucia 圣卢西亚岛 U7
staggering *a.* 难以置信的，令人震惊的 U12
stagnant *a.* 停滞的，不景气的 U7
stewardship *n.* 管理工作 U6

Strait of Gibraltar　直布罗陀海峡　U8
Strait of Hormuz　霍尔木兹海峡　U8
Strait of Malacca　马六甲海峡　U8
strand　*v.* 搁浅；陷入困境　U5
stranglehold　*n.* 束缚；压制；勒紧　U4
streamline　*v.* 把…做成流线型；使简单　U12
stretcher　*n.* 担架　U5
structural handicap　结构性障碍　U7
subduction　*n.* 消减，潜没，俯冲（作用）（指地壳的板块沉到另一板块之下）　U1
submarine　*a.* 水下的，海底的　U3
submarine mountains and valleys　洋底山脉与山谷　U1
subsidize　*v.* 以津贴补助　U12
substrate　*n.* 基质，底层　U3
suffocate　*v.*（使）窒息而死；阻碍　U12
suffocation　*n.* 窒息；闷死　U4
sulfide　*v.* 硫化物　U4
sulfur　*n.* 硫黄（色）　*v.* 用硫黄处理　U1
sulphide　*n.* 硫化物　U3
supranational　*a.* 超国家的，超民族的　U6
surge　*v.* 蜂拥而来；汹涌；起大浪　U12
surveillance　*n.* 盯梢，监督；[法] 管制，监视　U12
sustainability　*n.* 可持续性　U7
swamp　*n.* 湿地；沼泽（地）　U12

T t

tailings　*n.* 尾矿地；尾矿；尾材　U4
tally　*n.* 记录，测量　U5
tangle　*n.* 纠缠；混乱；争论　*v.* 缠结；使陷入　U12
tap　*v.* 开发，利用；轻击，轻叩　U9
tariff　*n.* 关税；关税表　*v.* 征收关税；定税率　U11
taxonomic classification　系统分类　U3
tectonic　*n.* 构造的　U3
tellurium　*n.* 碲　U3

temperature gradient 温度变化率 U1
territorial *a.* 领土的；区域的；陆地的；地球的 U12
territory *n.* 领地；版图；领域；[商] 势力范围 U11
the Balearic Islands 巴利阿里群岛 U7
the Canary Islands 加那利群岛 U7
the Caribbean Islands 加勒比群岛 U7
the Channel Islands 海峡群岛 U7
the East Pacific Rise 东太平洋（海隆）海岭 U1
the Galapagos Islands 加拉帕格斯群岛 U7
the Georges Bank 乔治沙洲 U2
the Grand Banks （纽芬兰）大浅滩 U2
the Malvinas Islands 马尔维纳斯群岛 U7
the mid-oceanic ridge system 大洋中脊（隆）系统 U1
the solar nebula 太阳星云 U1
the solar system 太阳系 U1
thrive *v.* 茁壮成长；兴盛 U4
tidal bay 潮汐湾 U9
tidal energy 潮汐能 U9
tidal flushing 潮水冲洗力 U9
tidal power plant 潮汐发电站 U9
tidal range 潮差，潮位变幅，潮汐变化范围 U9
tidal rhythm 潮汐韵律（层） U1
tidal stream 潮流，潮汐河流 U9
tide mill 潮磨；潮轮 U9
tilapia *n.* 罗非鱼 U11
timescale *n.* 时间表 [尺度]，时标，时间量程 U1
titanium *n.* 钛 U3
trace *n.* 痕迹，微量 U3
transform faults 转换断层 U1
transgenics *n.* 转基因 U11
transient *a.* 路过的，短暂的 U8
transit *n.* 经过，运输 U8
transoceanic *a.* 跨洋的，在海洋彼岸的 U4
transparent *a.* 透明的 U12

transshipment *n.* 转载 U10
trawl *v.* 用拖网捕鱼 U12
trawler *n.* 拖网渔船 U2
trawling *n.* 拖网捕捞 U2
treaty *n.* 条约；协议，协商；谈判 U12
trench *n.* 海沟 U1
triage *n.* 医疗类选法 U5
tribal *a.* 部落的，部族的 U6
trophic chain 营养链 U11
trove *n.* 发现物，收藏物 U3
tuna *n.* 金枪鱼 U2
turbidity *n.* 混浊度 U4
turbine *n.* 涡轮（机） U9
turbulence *n.* 湍流；动荡 U2

U u

uncover *v.* 揭露，发现 U6
underlie *v.* 构成…的基础（或起因） U6
underline *v.* 在…下面画线；强调 U11
underlying rock 基岩，下垫岩石，下伏岩石 U1
undermine *v.* 逐渐削弱；从根基处破坏 U12
UNEP（United Nations Environment Programme） 联合国环境规划署 U6
uniform *a.* 一样的；始终如一的 *v.*（使）规格一致；（使）均一 U10
unilateral *a.* 单边的 U8
unimpeded *a.* 不受阻碍的 U8
upstream *a.* 向上游的，逆流而上的 *adv.* 向上游，逆流地 U9
upwelling *n.* 上涌；上升流 U2
UV *abbr.* 紫外线（的）（= ultraviolet） U1

V v

vaccine *n.* 疫苗，痘苗 *a.* 疫苗的，痘苗的 U11
vector *n.* 航线；矢量，向量 *v.* 用无线电导航 U10
Venus *n.* 金星 U1
vertically *adv.* 垂直地；直立地；陡峭地 U10

vessel *n.* 容器；船 U5
vestige *n.* 遗迹，残余 U7
vie *v.* 竞争 U8
Virgin Islands 维尔京群岛 U7
volatiles *n.* 挥发物 U1
volcanic islands 火山岛 U1
volcanic rocks 火山岩 U1
volcanism *n.* 火山活动，火山作用，火山现象 U1

W w

waterwheel 水车；吊水机 U9
wave energy 波（浪）能；海浪能 U9
Western Samoa 西萨摩亚岛 U7
whiteleg shrimp（*Litopenaeus vannamei*） 凡纳滨对虾 U11
whopping *a.* 巨大的 U2
wildlife reserve 禁猎区，野生动物保护区 U7
wind energy 风能 U9
wreak havoc upon 肆虐；破坏 U4

Y y

yellowfin *n.* 黄鳍金枪鱼 U2
yellowtail flounder 美洲黄盖鲽 U2
yield *vt.* 生产 *vi.* 退让 *n.* 产量；收益；屈服 U11
yield-per-recruit 单位补充渔获量 U2

Z z

zebra mussel 斑马贻贝 U4
zinc *n.* 锌 U4
zooplankton *n.* 浮游动物 U4

Appendix 2　Keys to Exercises

附录 2　参考答案

Unit One　The History of the Oceans

第一单元　海洋演变史

Ⅳ. Exercise 1：Sentences in Focus　练习 1：单句翻译

1. Translate the following sentences into Chinese

（1）最重要的是，为什么地球上存留有液体水，而太阳系中的其他星球上似乎又没有呢?

（2）可能是由于太热，构成水圈和大气圈的水分以及其他挥发物难以在高浓度下存留，于是便形成了像金星、地球和火星这样的岩石行星。

（3）地球太热，地表水难以存留，于是便在沸腾中蒸发变成了灼热的大气层。

（4）由轻元素组成外部地壳，类似洋葱皮，它与火山岩一样是从地幔层剥离出来的。

（5）这种分离——把地球分为了若干层，而每一层随着从地核向外延伸其密度会变小——发生得非常迅速。根据地质标准，这是在地球形成后的头一亿年内发生的。

（6）这一理论认为某些团块物掺加进了地球，但是，这一撞击使得大部分物质（包括地球上的大量物质）飞离，被困在地球的引力场中。

（7）月球对地球的引力使得地球略有伸缩，并导致海洋出现有节奏的潮汐运动。

（8）自外而内模型提出，地球在形成时没有水的存在，而是陨石或彗星撞击早期地球时形成的挥发物带来了水。

（9）另一方面，陨星似乎又是我们这个行星的建构模块，因为已经发现它们含有类似早期地球的组成成分。

（10）地球地幔层的含水量估计是海洋的 3～6 倍之多，所以，极有可能我们的地表水是来自于地球内部。

（11）地球科学家认为，在太阳系形成的头几亿年里，当地球分成地核、地幔和地壳时，这颗行星便开始通过火山活动排放气体（通过“打嗝”喷出挥发物）。

（12）今天，火山运动还在持续不断地向大气圈和水圈释放大量相同的气体——设想一下那些从活火山喷涌而出的白色羽流吧——不过，与早期地球释放出的巨大热量相比，其水平要低得多。

（13）尽管我们地球上的海洋和大气层已演变成一个可孕育生命的系统，但我们相邻的行星的境遇却不是那么理想。

（14）火星的两极被认为含有延伸到地下的冰冻水和二氧化碳冰帽。最近漫游者探险的证据表明曾经有液态水在这个红色的星球表面存在过。

（15）所有这些因素促成了一个近乎完美的动态系统，它不仅使大洋存在于这颗行星之上，而且保护着栖居于此的生物体。

2. Translate the following sentences into English

(1) To physical science that has denatured reality, the world is a material place of molecules and atoms, of solids, liquids and gases, of atmosphere, hydrosphere, lithosphere and biosphere.

(2) Pangaea began to break apart about 225 million years ago due to geological processes related to plate tectonics, which eventually spread Earth's landmasses into the continents we see today.

(3) Continental crust averages some 25 miles (40 kilometers) thick, although it can be thinner or thicker in some areas. Oceanic crust is usually only about 5 miles (8 kilometers).

(4) "Volcanism is important because it represents the pulse of the planets," he added.

(5) A recent analysis of lunar rocks reveals that they have the same concentration of water as the Earth's upper mantle, the layer of near-molten rock just beneath the crust.

(6) Moreover, even when the crabs were taken from the beach and put back in the dark, they continued their tidal rhythm.

(7) Gravity never disappears completely but to get weak. And every object with mass, including you and me, has what's known as gravitational attraction.

(8) But geophysicists had thought that great subduction-zone earthquakes

happened only where younger oceanic crust scrapes its way into the mantle.

(9) Two weeks ago, the US astronomers announced that they had discovered the golden planet orbiting another star, of which the size and temperature are suitable for the extraterrestrial life to live in.

(10) The researchers warned that if another earthquake hit the region, Sumatra island coast near Padang City will sink tens of centimeters.

(11) The Satellite photos taken three days ago show that the famous north-west and north-east waterways are clear because the ice are beginning to melt, which make it possible for the ships to sail in waters near the North Pole ice cap.

(12) The island nation is actually a mountain top of the mid-Atlantic ridge, which is one of the longest mountain ranges of the world.

(13) In 1991, one of large volcanoes on the East Pacific Rise erupted. The lava almost swept across the sea floor as well as blocked the hydrothermal vents, thus, killing the marine creatures near the vent.

(14) All places along the Pacific trench (the Pacific ring of fire), should be evacuated very soon, because earthquakes and volcanoes will erupt.

(15) And also, if the ocean water can be removed, the sea floor, the wide valley, rugged mountains and submarine rivers would present an unimaginable picture.

Ⅴ. Exercise 2: Passages in Focus 练习 2：短文翻译

1. Translate the following passages into Chinese

海洋不仅是海水刚好覆盖陆地的那些区域。从地质学角度来讲，洋底与陆地是截然不同的。洋底处于一种永不停息的诞生和毁灭的轮替之中，从而形成海洋，并在很大程度上控制着大陆地质及地质演变。海水下发生的地质过程不仅影响海洋生物，也对陆地产生影响。洋盆形成的过程十分缓慢，需经过数千万年的时间。根据这一时间量表，人的一生不过是眨眼工夫之事。固体岩石的移动就像液体流动一样，整个陆地穿行于地球表面，平原变成了高山。为弄清洋底情况，我们必须学会接受不为人熟知的地质时期的观点。地质过程对海洋生物学至关重要。一些栖息地，或生物体生存的空间，都是地质过程作用的直接结果。海岸线的形状，海水的深度，洋底是淤泥、沙地还是岩石，以及海洋栖息地的诸多其他特征都取决于这一地质过程。生物的地质史也称古生物学。

二战后的一些年里，声呐的出现使人类第一次可以对洋底的大片区域展开详细的调查。这些调查最终导致了大洋中脊系统的发现。这是一系列绵延40 000英里像棒球接缝一样环绕地球的洋底火山山脉和山谷。地球上最大的地质特质就是大洋中脊系统。每隔一段时间，大洋中脊就会被地壳的裂缝，即转换断层，向一边或另一边推移。有时候，洋底中脊山脉会升高，冲破洋面，形成诸如冰岛、亚速尔群岛等岛屿。

大洋中脊处于大西洋的部分，称为大西洋中脊，纵贯大西洋中部，紧紧地与相对的海岸线的曲线保持一致。洋脊在印度洋形成一倒立的 Y 形，然后沿着太平洋东面向上延伸。太平洋东面洋脊的主要部分称为“东太平洋海隆”。调查还表明，洋底存在一系列的深坳陷区域，称为海沟。海沟在太平洋尤其普遍。

海底磁异常的发现，再加上一些其他证据，最终使人们弄清了板块的构造。地球表面可分成若干板块。这些板块，由地壳和地幔的顶端部分组成，形成了岩石圈。这些板块厚度约 100 km。新的岩石圈产生的同时，古老的岩石圈会在其他地方毁灭。否则，地球就不得不持续地膨胀，来为新生的岩石圈让出空间。岩石圈在海沟处毁灭。当两个板块相撞击，其中一个板块浸没于另一板块之下，然后回移向下挤入地幔时，就会形成海沟。这种板块挤入地幔的下向运动称为俯冲。由于俯冲发生在海沟处，海沟常被称为俯冲带。俯冲这一过程还会导致水下地震和火山的产生。火山可以源于洋底，形成火山岛链。

现在，我们了解到地球表面经历了剧烈的变化。大陆因洋底运动而长距离地运移，从而使洋盆在大小与形状上也产生了变化。事实上，新的大洋已经诞生了。了解板块构造的过程使科学家们可以对这些变化的大部分历史进行重新构建。比如说，科学家们就已经发现，大陆曾经是连成一片的一个单一的超大陆，称为盘古大陆，1.8 亿年前才开始分裂开来。自此，大陆才漂移到目前所处的位置。

2. Translate the following passages into English

The ocean is traditionally classified into four large basins. The Pacific is the deepest and largest, almost as large as all the others combined. The Atlantic Ocean is a little larger than the Indian Ocean, but the two are similar in average depth. The Arctic is the smallest and shallowest. Connected or marginal to the main ocean basins are various shallow seas, such as the Mediterranean Sea, the Gulf of Mexico and the South China Sea.

The earth is composed of three main layers: the iron-rich core, the semiplastic mantle and the thin outer crust. The crust is the most familiar layer of the earth. Compared to the deeper layers it is extremely thin, like a rigid skin floating on top of the mantle. The composition and characteristics of the crust differ greatly between the oceans and the continents.

Though we usually treat the oceans as four separate entities, they are actually interconnected. This can be seen most easily by looking at a map of the world as seen from the South Pole. From this view, it is clear that the Pacific, the Atlantic and Indian oceans are large branches of one vast ocean system. The connections among the major basins allow seawater, materials, and some organisms to move from one ocean to another. Because the oceans are actually one great interconnected system, oceanographers often speak of a single world ocean. They also refer to the continuous body of water that surrounds Antarctic as the Southern Ocean

The geological distinction between ocean and continents is caused by the physical and chemical differences in the rocks themselves, rather than whether or not the rocks happen to be covered with water. The part of earth covered with water, the ocean, is covered because of the nature of the underlying rock.

Oceanic crustal rocks, which make up the sea floor, consist of minerals collectively called basalt that have a dark color. Most continental rocks are of general type called granite, which has a different mineral composition than basalt and is generally lighter in color. Ocean crust is denser than continental crust, though both are less dense than the underlying mantle. The continents can be thought of as thick blocks of crust floating on the mantle, much as icebergs float on water. Oceanic crust floats on the mantle too, but because it is denser it doesn't float as high as continental crust. This is why the continents lie high and dry above sea level and oceanic crust lies below sea level and is covered by water. Oceanic rocks and continental crust also differ in geological age. The oldest oceanic rocks is less 200 million years old, quite young by geological standards. Continental rocks, on the other hand, can be very old, as old as 3. 8 billion years.

Unit Two　Marine Biological Resources

第二单元　海洋生物资源

Ⅳ. Exercise 1：Sentences in Focus　练习 1：单句翻译

1. Translate the following sentences into Chinese

(1) 人类在海洋中捕鱼已经有数千年的历史，海洋是人类社会不可或缺的一部分。

(2) 今天，全世界大约 16%的蛋白质来自渔业，在发展中国家，这个比例还要更高。

(3)“渔业”一词指所有的海洋捕捞活动，无论是为了商业捕捞、休闲捕捞，还是为了获得观赏鱼或鱼油。

(4) 由于大陆架上鱼类资源相对丰富，渔业通常是指海洋渔业，而不是淡水渔业。

(5) 不同年份的渔获量会有差异，但也许是因为过度捕捞以及经济和管理方式，其似乎稳定在每年约 8 800 万吨。

(6) 捕鱼方式有很多，包括单人撒网、巨大的拖网、围网、流刺网、手钓、延绳钓、刺网和潜水捕捞。

(7) 渔获物并非总是被用作食物。实际上，有大约 40%的鱼被用于其他用途，如做成鱼粉用来人工养殖鱼类。

(8) 由于自然原因和人类发展，海洋里能够捕捞的鱼的数量在不断变化。

(9) 由于人口不断增加，海洋已经被过度捕捞，导致对于世界经济和社会至关重要的鱼类数量减少。

(10) 科学家常要承担起渔业管理者的角色，而且必须控制海洋渔业捕捞的数量，而由于鱼类种群不断减少，这样的角色一定不受那些以捕捞为生的人欢迎。

(11) 海洋里有多少鱼类？这个问题很复杂，但是如果根据鱼类种群而不是个体来计算，可以让这一问题变得简单。

(12) 世代群组死亡率和鱼的种类密切相关，因为不同种类的鱼之间自然死亡率会有差异。

(13) 根据生产理论，当鱼的数量不会让环境不堪重负，种群的基因多样性得以维持时，鱼的产量将会是最高的。

(14) 就连海洋湍流也会影响掠食者和被掠食者之间的关系，当海水被搅动起来时，两者会有更多相遇的机会。

（15）被丢弃的鱼类和其他海洋生物被称为副渔获物，这是不可持续且惯常的捕捞方式所带来的不幸副作用，会打乱生态系统，在海水中留下大量的无机物。

2. Translate the following sentences into English

（1）The biomass depends on the amount of carbon fixed by green plants and other producers.

（2）In captivity，gorillas have displayed significant intelligence and have learned simple human sign language.

（3）The widespread use of large-scale driftnets is a destructive fishing practice that poses a threat to living marine resources of the world's oceans.

（4）Estuaries，where freshwater rivers meet the salty sea，are popular destinations for fishing，boating，birding and hiking.

（5）Longlining can cause many issues，such as the killing of many other marine animals while seeking certain commercial fish.

（6）One of the most significant issues affecting marine fisheries management today is the mortality of fish that are discarded after capture or that escape from fishing gear.

（7）Wherever there is fishing，there is bycatch—the incidental capture of non-target species such as dolphins，marine turtles and seabirds.

（8）Ornamental plants and animals that are accidentally or deliberately released into the wild can establish reproducing populations，often with disastrous impacts on native ecology.

（9）In an ecosystem，predation is a biological interaction where a predator feeds on its prey.

（10）The western and central Pacific Ocean is one of the main fishing grounds of tuna purse seine fishery in the world with the main target species of skipjack，yellowfin and bigeye tuna.

（11）Juvenile animals sometimes look very different from the adult form，particularly in colour.

（12）Even a minor fluctuation in the water temperature can affect the fish.

（13）Often associated with low temperatures，the function of hibernation is to conserve energy during a period when sufficient food is unavailable.

（14）If you are having trouble gaining weight or muscle，that is usually the first sign that you may have a fast metabolism.

(15) Upwelling currents commonly exist everywhere in the ocean, driven by wind and the landscape, flowing from sea bottom bringing nutrition to sea surface, providing resources for sustainable development in the ocean.

Ⅴ. Exercise 2: Passages in Focus 练习 2：短文翻译

1. Translate the following passages into Chinese

在过去的 50 年里，全球渔业产量稳步增长，食用鱼的供应以平均每年 3.2%的速度增加，超过了世界人口每年 1.6%的增长速度。世界人均鱼类表面消费量从 20 世纪 60 年代的 9.9 千克增加到了 2012 年的 19.2 千克。这一显著增长是由多种因素共同驱动的，其中包括人口增长、收入增加和城市化的发展，而渔业产量的大幅增长和更加高效的分销渠道也起了助推作用。更多的人能够吃到鱼，在这方面，中国功劳最大，因为中国的渔业产量、尤其是水产养殖业的产量有了大幅度的增长。从 1990 年到 2010 年，其人均鱼类表面消费量平均每年增长率为 6%，到 2010 年，人均消费已经约为 35.1 千克。2010 年，世界上其他地方人均鱼类的年供应量约为 15.4 千克（20 世纪 60 年代为 11.4 千克，90 年代为 13.5 千克）。

虽然发展中地区和低收入缺粮国人均每年鱼类表面消费量都有了很大的增长，前者是从 1961 年的 5.2 千克增加到了 2010 年的 17.8 千克，后者是从 4.9 千克增加到了 10.9 千克，但是发达地区的消费水平依然更高，不过两者之间的差距正在缩小。发达国家所消费的鱼类份额相当大，并不断增加，需要进口，这一方面是因为稳定的需求，另一方面是因为国内渔业产量的下降。在发展中国家，鱼类的消费往往是基于一些可以买得到产品，具有地方性和季节性，要靠供给来驱动产品链。但是，随着国内收入和财富的增长，以及渔业进口的增加，新兴经济体的消费者有了更多的鱼类可以选择。

一份 150 克的鱼肉可以提供一个成人一天所需要的蛋白质的 50%～60%。2010 年，鱼类占全球人口动物蛋白摄入量的 16.7%，占蛋白摄入总量的 6.5%。此外，鱼类为全球 29 亿人口提供了动物蛋白摄入量的几乎 20%，为 43 亿人口提供了动物蛋白摄入量的大约 15%。在有些人口稠密、蛋白摄入量可能较低的国家，鱼类蛋白是至关重要的营养成分。

2011 年，全球渔获物产量为 9 370 万吨，是有史以来第二高的一年（仅次于 1996 年的 9 380 万吨）。此外，除了鳀鱼的产量之外，2012 年是又一个产量最高的年份，为 8 660 万吨。但是，这些数据所体现的是对过去报告的总体稳定形势的延续。2011 年和 2012 年全球海洋渔业的产量分别为 8 260 万吨和

7 970万吨。在这几年中，有 18 个国家（11 个在亚洲）平均每年捕获量超过了 100 万吨，占全球海洋捕获量 76%以上。

西北和中西太平洋是捕获量最高的区域，并且还在增长。东南太平洋的产量总是受到气候变化的很大影响。在东北太平洋，2012 年的总捕获量和 2003 年相同。2012 年，印度洋的捕获量保持了长期以来持续增长的势头。在 2007 年至 2009 年之间，海盗活动对西印度洋的渔业产生了不利影响，但是此后金枪鱼捕获量已经恢复。在 2011 年和 2012 年，北大西洋、地中海和黑海的渔获产量再次下降。西南和东南大西洋的产量最近一直在恢复。

2. Translate the following passages into English

World aquaculture production continues to grow, albeit at a slowing rate. According to the latest available statistics collected globally by FAO, world aquaculture production attained another all-time high of 90.4 million tonnes (live weight equivalent) in 2012 (US$144.4 billion), including 66.6 million tonnes of food fish (US$137.7 billion) and 23.8 million tonnes of aquatic algae (mostly seaweeds, US$6.4 billion).

In addition, some countries also reported collectively the production of 22,400 tonnes of non-food products (US$222.4 million), such as pearls and seashells for ornamental and decorative uses. For this analysis, the term "food fish" includes finfishes, crustaceans, molluscs, amphibians, freshwater turtles and other aquatic animals (such as sea cucumbers, sea urchins, sea squirts and edible jellyfish) produced for the intended use as food for human consumption. At the time of writing, some countries (including major producers such as China and the Philippines) had released their provisional or final official aquaculture statistics for 2013. According to the latest information, FAO estimates that world food fish aquaculture production rose by 5.8 percent to 70.5 million tonnes in 2013, with production of farmed aquatic plants (including mostly seaweeds) being estimated at 26.1 million tonnes. In 2013, China alone produced 43.5 million tonnes of food fish and 13.5 million tonnes of aquatic algae.

The total value of global aquaculture has probably been overstated owing to factors such as some countries reporting retail, product or export prices instead of prices at first sale. Nonetheless, when used at aggregated levels, the value data are useful in showing the development trend and for comparison of the relative importance of economic benefit among different types of aquaculture and different groups of farmed aquatic species.

The global trend of aquaculture development gaining importance in total fish supply has remained uninterrupted. Farmed food fish contributed a record 42. 2 percent of the total 158 million tonnes of fish produced by capture fisheries (including for nonfood uses) and aquaculture in 2012. This compares with just 13. 4 percent in 1990 and 25. 7 percent in 2000. Asia as a whole has been producing more farmed fish than wild catch since 2008, and its aquaculture share in total production reached 54 percent in 2012, with Europe at 18 percent and other continents at less than 15 percent.

The overall growth in aquaculture production remains relatively strong owing to the increasing demand for food fish among most producing countries. However, aquaculture output by some major producers, most notably the United States of America, Spain, France, Italy, Japan and the Republic of Korea, has fallen in recent years. A decline in finfish production is common to all these countries, while mollusc production has also decreased in some of them. The availability of fish imported from other countries where production costs are relatively low is seen as a major reason for such production falls. The resulting fish supply gap in the aforementioned countries has been one of the drivers encouraging production expansion in other countries with a strong focus on export-oriented species.

World food fish aquaculture production expanded at an average annual rate of 6. 2 percent in the period 2000 - 2012, more slowly than in the periods 1980 - 1990 (10. 8 percent) and 1990 - 2000 (9. 5 percent) . Between 1980 and 2012, world aquaculture production volume increased at an average rate of 8. 6 percent per year. World food fish aquaculture production more than doubled from 32. 4 million tonnes in 2000 to 66. 6 million tonnes in 2012.

Unit Three Marine Mineral Resources

第三单元 海洋矿产资源

Ⅳ. Exercise 1: Sentences in Focus 练习 1：单句翻译

1. Translate the following sentences into Chinese

(1) 自从 1994 年生效以来，这一重要公约已经成为签约国利用领海之外海底海洋资源合法权利的基础。

(2) 当前兴趣的复苏是因为资源价格的急剧增长，以及随之而起的采矿业利润的上升，对于像中国和印度这样经济增长比较强劲的国家来说，情况尤其如此，它们从世界市场上购买大量金属。

(3) 锰结核是块状的矿物质，其大小在土豆和莴苣之间不等，分布于深海的巨大区域，每平方米多达 75 千克。

(4) 这些化学元素是从海水中沉淀而成的，或者是源自于底层沉积物的孔隙水。

(5) 此外，还有其他重要元素的痕迹，如铂或碲，这些元素对于各种高科技产业非常重要。

(6) 实际的开采过程并不存在技术上的重大问题，因为结核可以被轻而易举地从海底表面采集。

(7) 大量的沉积物、海水和无数的有机体会和结核一起被挖起来，从而对深海栖息地造成巨大的破坏。

(8) 开采钴结壳还会对底栖生物造成很大的影响，因此在开采之前必须要对其环境影响进行研究。

(9) 这些埋藏的大型硫化矿，形成于海底的板块边界，在这里，由于火山活动和海水的互动，地壳的岩石和海洋之间会发生热量和元素的交换。

(10) 长期以来，人们一直认为大量具有开采潜力的硫化物仅仅形成于大洋中脊，因为这里的火山活动和热量的产生尤其剧烈。

(11) 东太平洋海隆和大西洋中部的黑烟柱产生的硫化物主要是富含铁的硫化合物，这些不值得考虑去进行深海开采，而太平洋西南部的埋藏物却包含大量的铜、锌和黄金。

(12) 当前的开采方案主要是针对冷却的、不活跃的大型硫化矿存象，这里生物栖息稀少。

(13) 开采作业本来计划今年开始，但是由于当前的经济衰退，虽然黄金的价格相对较高，但主要的金属和采矿公司营业额下降，而且工程在短期内被延期。

(14) 虽然海底有大量的资源，但是只有当有足够的需求并且金属价格相对很高时，海床开采才能和当前陆上巨大可采矿藏一较高下。

(15) 要把外壳从基质分离出来，还有很大的技术难度，加上海底表面崎岖不平所带来的困难，进一步降低了当前钴结壳的经济潜力。

2. Translate the following sentences into English

(1) More-recent work indicated massive stars could form by accretion after all, but researchers had not actually observed it.

(2) Some researchers have argued that life could have begun around deep-sea hydrothermal vents.

(3) A major significant failure of the resurgence of market-oriented policies was a failure to reform the monetary system.

(4) The disease appears to be endemic in the country, and surveillance and control campaigns have so far not succeeded in interrupting virus transmission between provinces.

(5) The fact that Google is sitting on an information treasure trove is one thing. Knowing how to take advantage of that fact is another.

(6) Sediment deposit leads channel and river mouth to deposit, and discharge capacity descend, so the dangers of flood disaster is more and more serious.

(7) A gas in solution diffuses from a region of greater to one of less concentration.

(8) The main commitment made by every signatory state is to develop a national strategy (or action plan) to manage and protect their own biodiversity.

(9) Indonesia is located on the edges of tectonic plates, specifically the Pacific, Eurasian and Australian. As such, it is frequently hit by earthquakes.

(10) Deforestation can also change soil dynamics and increase erosion, both of which can release more carbon into the atmosphere.

(11) Differences in the concentration of trace elements are related to the climatic conditions and land use of the area.

(12) The team looked at fossils of so-called benthic organisms, such as starfish, clams and corals that live on the seabed.

(13) Before you undertake such transactions, you should familiarize yourself with applicable rules and attendant risks.

(14) The mud and gas seep through faults, eventually erupting at the surface as a mud volcano.

(15) Coral reefs are some of the world's richest ecosystems, supporting a diverse array of marine life, including coral, fish and mollusks.

Ⅴ. Exercise 2: Passages in Focus 练习 2：短文翻译

1. Translate the following passages into Chinese

一旦在海洋中发现一种矿藏，下面必须要解决的问题就是这种矿物是否有用（能够用来交易），以及是否有技术能够将这种有价值的产品开采出来。这是一个序列过程的起点，在此过程中，人们会逐步尝试确定上述矿产的储量。这一过程的第一步是对广大的海底区域进行地方性的评估，了解其矿藏，以便选出最有开采前景的区域，进行进一步的勘查，这一阶段通常被称为找矿阶段。下一阶段是勘探，目的是确定矿床，了解其特征，确定是否有现成的技术还是要开发新技术，以便从中获得有价值的产品。如果有现成的、经过测试的技术，就会很快进入试开采阶段，以确定开采的经济前景。对于大陆架上的海洋矿产来说，开采的传统已经相对比较悠久，从发现到生产的过程直截了当。技术的开发通常集中在技术适应方面。

然而，国际海底区域矿产资源开采会面临一些挑战，包括高昂的勘探成本（时间和经济），同时，这一区域还没有一种矿物得到开采，因为这些矿物（多金属结核、海底多金属大型硫化矿以及钴锰铁结壳）的有关规章制度和程序正在建立之中，必须要借助应用科学，开发出成本更低的勘探方法。

对于今后的开采，尤其是多金属结核的开采，在国际海底管理局保护海洋环境的工作中，科技正发挥着重要的作用。例如，在克拉里昂克利帕顿断裂带，海底管理局已经和 6 个实体签署了勘探合同，当前正在进行的项目鼓励环境数据收集的标准化，包括对该区域的动物群系统分类，确定这里的物种范围等。这些项目的结果有望帮助海底管理局和签约者控制结核采矿业的影响。但是，主要的挑战依然是采矿技术。在这一方面，如果没有新的采矿技术取代二十世纪七八十年代的配置，对于海底多金属结核的开采何时开始的讨论，将继续聚焦于其中所包含的有价元素价格的增长。

2. Translate the following passages into English

A scientific revolution in our understanding of the way the Earth works started in the 1960's that significantly expanded our knowledge of marine minerals. The scientific revolution entailed a major change in viewing the ocean basins and continents. Before the scientific revolution, the ocean basins were viewed as big bathtubs that passively contained the oceans. The continents and ocean basins were viewed as permanent features that had remained in their present positions through most of Earth's history.

The marine mineral provisions of the UNCLOS recognized those non-fuel marine mineral deposits that were derived from erosion of land and carried into the ocean in particulate or dissolved form by rivers. These minerals comprised heavy metal deposits, gemstones (especially diamonds), sand and gravel deposited in sediments of continental margins, phosphorites deposited on hard rock substrates of continental margins, and polymetallic nodules precipitated on the floor of the deep ocean from metals dissolved in seawater.

The scientific revolution, based on a theory called plate tectonics, changed our view of ocean basins from big bathtubs to dynamic features that open and close on a time scale of tens to hundreds of millions of years with concomitant movement of the land areas known as continental drift. The scientific revolution recognized the ocean basins as sources of types of non-fuel mineral deposits. These newly recognized types of marine mineral resources include polymetallic massive sulphides containing copper, iron, zinc, silver, gold and other metals in varying amounts. Polymetallic sulphides deposits are concentrated over thousands of years by seafloor hot springs at sites along an active submerged volcanic mountain range that extends through all the ocean basins of the world.

Another newly recognized type of marine mineral resource is cobalt crusts that are precipitated over millions of years on the submerged flanks of inactive underwater volcanoes from metals dissolved in seawater derived from input of metals by both rivers and seafloor hot springs. None of these newly recognized types of marine mineral deposits are renewable resources, as they all require thousands to millions of years to accumulate in economically interesting grades and tonnages. They are resources for the future with no present production.

We are still at an early stage in exploration of the oceans with only a few percent of the seafloor known in detail and even less known about what lies beneath the seafloor, so new discoveries will continue to be made.

Unit Four Marine Pollution

第四单元 海洋污染

Ⅳ. Exercise 1: Sentences in Focus 练习 1：单句翻译

1. Translate the following sentences into Chinese

（1）营养物污染是水污染的一种形式，指的是由营养物质过量输入而导致的污染。它是表层海水富营养化的一个主要原因。过量的营养物通常为氮或磷，它们刺激着藻类在海面的生长。

（2）微小颗粒物上黏附着很多具有潜在毒性的化学物质，这些颗粒物常被浮游生物和海底生物吸食，其中的多数生物是食碎屑动物或滤食动物。

(3) 有毒金属也会被带入海洋食物网，它们能改变海洋生物的机体组织物质和生物化学构成，影响它们的行为方式和繁殖模式，抑制它们的生长。

(4) 体内汞浓度最高的滤食性桡足类动物不是出现在这些河流的河口处，而是出现在河口以南 70 英里，更靠近亚特兰大市的区域。这是因为河水是靠着海岸流动的，并且毒素被浮游生物吸收也需要一些时日。

(5) 然而，采矿过程中排入海洋的有些矿物会导致问题的出现，如，铜，这种常见的工业污染物会扰乱珊瑚虫的生命周期，影响其生长。

(6) 当塑料碎片太大或者纠缠在一起时，是很难排出的；这样一来，它们就可能永久地留在这些动物的消化道里。

(7) 在海洋环境下，被光致降解的塑料虽然分解成了更小的碎片，但它仍是聚合物——即便是已经降解至分子形态。

(8) 这些废弃的渔网，俗称“夺命网”，能缠住鱼、海豚、鲨鱼、儒艮、螃蟹和其他生物，限制它们的活动，造成饥饿、裂伤和感染；而对于那些需要返回洋面呼吸的生物来说，则意味着窒息而亡。

(9) 营养丰富的海水会引起沿海区域的肉质藻类和浮游植物疯长，这种现象俗称藻华；因为几近消耗了所有的氧气，藻华现象可能造成该区域缺氧。

(10) 到了暖季，沿着亚热带高压脊南部边缘地带游走的撒哈拉尘沙，就会进入加勒比海和佛罗里达区域，这是因为该高压脊沿北向形成，贯穿了亚热带大西洋区域。

(11) 海洋酸化的潜在后果尚未被人类全面认知，但是已经有人在担心由碳酸钙组成的结构会变得易于溶解，从而影响到珊瑚和贝类动物贝壳的形成。

(12) 虽然油轮失事会引发大范围的头条新闻，但是世界海域内的多数石油污染来自于一些更小的源头，例如，油轮返航时从油罐排放的压载水，渗漏的输油管，或是直接排放到下水管道中的发动机油。

(13) 据悉，一个因单一物种入侵而危害到生态环境的最严重的实例是由一种表面上看似无害的水母引起的。

(14) 然而，专家们确信的是，局部海床的移动会扰乱深海底层的环境，导致水柱毒性增大，还会引发更多来自尾矿的沉淀物羽流。

(15) 漂浮的颗粒物会加深海水的混浊度，使之变得模糊不清，还会堵塞滤食性底栖生物机体内的过滤装置。

2. Translate the following sentences into English

(1) Plastic bags do not biodegrade. They photodegrade breaking down into smaller and smaller toxic bits constantly contaminating soil and waterways

and entering the food web when animals accidentally ingest.

(2) The results showed that the active compositions are of water-soluble protein, and susceptible/ vulnerable to acidity, heat, and can be restricted by ethanol.

(3) Not only that, but matter itself will have fallen apart: protons, long thought to be utterly stable, may disintegrate after about 1,039 years.

(4) Afloat on a boat or flotsam, they will be dragged inland with flow until a reverse slosh begins, the water flowing back into its bed.

(5) More and more engineering (such as water-electrolyte, mine, nucleus debris burial and solute moving) are built on or in rock mass, then the problems of ground water which plays an important role of the engineering security, are putting in front of us.

(6) They replicate in their host cells so prodigiously and stream out into their surroundings so continuously that if you collected all the viral flotsam afloat in the world's oceans, the combined tonnage would outweigh that of all the blue whales.

(7) Turning crop wastes and other biomass into charcoal and spreading it on tropical soils can sequester carbon and boost crop productivity.

(8) Ion concentration analysis and turbidity analysis are used in this research to study the static compatibility of natural sea water and stratum water.

(9) The overall impact of the greenhouse effect is still unknowable, but it directly jeopardizes our ecosystem by causing the two-pole ice melting and sea level rising which are attributed to the global warming.

(10) The problem came in last March when a mutation of influenza virus showed up that scientists knew would wreak havoc upon the whole country, but it was too late.

(11) The Blue Revolution of freshwater aquaculture and mariculture is growing exponentially in that region.

(12) The micro-algae thrive on carbon dioxide, producing food for livestock as well as biofuels and materials for plastics.

(13) It plays an increasingly important part in meteorology for cloud, precipitation, hail and thunderstorm detection as well as the navigation of aircraft and ships.

(14) Today, people pay more and more attention to the environment problem and sustainable development of economy, environment and society, since ecosystem has been deteriorated and resources has been depletive.

(15) To ensure mass health and environmental safety, we launched the fourth annual nationwide campaign to punish enterprises that illegally discharge pollutants.

Ⅴ. Exercise 2: Passages in Focus　练习 2：短文翻译

1. Translate the following passages into Chinese

富营养化是沿海水域的一种常见现象。世界资源研究所已确认 375 个缺氧沿海地带，它们分布于世界各地，主要集中在西欧沿海地区、美国东部和南部沿海以及东亚，尤其是日本。跟淡水环境不同，海水中更常见的关键营养元素是氮；因此，氮含量在认识海水富营养化问题上起着更重要的作用。

一个解决江河入海口处富营养化问题的建议是恢复此处的贝类种群，如牡蛎。牡蛎礁能使水体脱氮，过滤掉悬浮物，从而减少或减轻有害藻华或缺氧状况发生的可能性或程度。贝类的滤食活动能控制浮游植物的密度，吸收营养素，所以被认为是有利于水质的。通过贝类养殖，过剩营养物质可以从生态系统中移除，或被掩埋在沉积物中，或通过反硝化作用流失。在瑞典，科学家们已经进行了一些基础性研究工作，以验证这种通过贝类养殖来改善海水水质的想法。他们养殖的是贻贝。

水体营养水平提升的一个典型后果是“藻华”或“赤潮”的发生，即浮游植物的急剧增多。科学家们更倾向于称之为“有害藻华”，而不是“赤潮”，因为藻华并不总是红色的，且其产生也跟潮汐无关。在海洋中，频繁出现的有害藻华会导致鱼类和海洋哺乳动物死亡；当其靠近海滨爆发时，则会引起人类和一些家畜的呼吸道疾病。美国所有沿海州都经历过有害藻华，这已引起了该国的关注，因为它们不仅影响到人与海洋生态系统的健康，还影响到了该国的经济“健康”——特别是在劳动收入主要依赖于渔业和旅游业的沿海地区。在美国，沿海藻华事件已经造成的经济影响估计每年至少达 8 200 万美元。

气候变化和日益严重的营养污染会导致有害藻华更加频发，甚至有可能在以前没有发生过的地方出现。所以重要的是，应尽可能地了解它们形成的方式和原因以及爆发的地点，以便减少其危害。研究表明，当风向和水流有利时，许多藻类就会茂盛生长。其他情况下的有害藻华则可能直接跟“过量喂食”有关。当来自陆地的各种养分流入海湾、河流和大海，并逐步累积到某种程度时，就会出现环境中正常存在的藻类被“过量喂食”的情况。某些有害藻华会发生在某些自然现象之后，如缓慢的水循环，异常的高水温，以及诸如飓风、洪水和干旱的极端天气之后。虽然许多沿海国家都经历过有害藻华，但是生长在不同地方的不同藻类会引起不同的问题。还有一些其他因素，如海岸的构造、径流、海洋学现象以及水里的其他生物，也可以改变有害藻华的影响范围和严重程度。

2. Translate the following passages into English

Oceans and coastal ecosystems play an important role in the global carbon cycle and have removed about 25% of the carbon dioxide emitted by human activities between 2000 and 2007 and about half the anthropogenic CO_2 released since the start of the industrial revolution. Rising ocean temperatures and ocean acidification means that the capacity of the ocean carbon sink will gradually get weaker, which has given rise to global concerns.

A report from NOAA scientists published in the journal *Science* in May 2008 found that large amounts of relatively acidified water are upwelling to within four miles of the Pacific continental shelf area of North America. This area is a critical zone where most local marine life lives or is born. While the paper dealt only with areas from Vancouver to northern California, other continental shelf areas may be experiencing similar effects.

A related issue is the methane clathrate reservoirs found under sediments on the ocean floors. These trap large amounts of the greenhouse gasmethane, which ocean warming has the potential to release. In 2004 the global inventory of ocean methane clathrates was estimated to occupy between one and five million cubic kilometers. If all these clathrates were to be spread uniformly across the ocean floor, this would translate to a thickness between three and fourteen meters. This estimate corresponds to 500 - 2,500 gigatonnes carbon, and can be compared with the 5,000 gigatonnes carbon estimated for all other fossil fuel reserves.

Apart from acidification, there exist some other types of pollution in the ocean environment, such as noise pollution. Marine life can be very susceptible to noise or the sound pollution from sources such as passing ships, oil exploration, seismic surveys, and naval low-frequency active sonar. Sound travels more rapidly and over larger distances in the sea than in the atmosphere. Marine animals, such as cetaceans, often have weak eyesight, and live in a world largely defined by acoustic information. This applies also to many deeper sea fish, who live in a world of darkness. Between 1950 and 1975, ambient noise at one location in the Pacific Ocean increased by about ten decibels (that is a tenfold increase in intensity).

Noise also makes species communicate louder, which is called the Lombard effect. Whale songs are longer when submarine-detectors are on. If creatures don't "speak" loud enough, their voice can be masked by anthropogenic sounds. These unheard voices might be warnings or finding of prey. When one species begins speaking louder, it will mask other species voices, causing the whole ecosystem to eventually speak louder.

Unit Five　Marine Disaster

第五单元　海洋灾难

Ⅳ. Exercise 1: Sentences in Focus　练习 1：单句翻译

1. Translate the following sentences into Chinese

(1) 什么是灾难？灾难是由自然因素引起，或者是人为造成的事件；灾难具有巨大破坏力，需要相关部门做出超常的反应。

(2) 需要记住的是，没有任何方案可以适用于所有情况，也没有任何灾难处置方案永远是一成不变的。

(3) 应对重大伤亡海难事故，了解海难现场管理的基本原则，具备常识认知、交际技能和良好的航海技术都是非常重要的。

(4) 总的来说，海上环境可能会使救援应对更加困难，因为除了需要照顾和撤离幸存者，海上紧急救援还常常涉及层出不穷的危险和错综复杂的情况。

(5) 遇难船只的船长是对其乘客和船员的最终负责人。这包括及时撤离船上人员（如果有必要），与搜救部门协力展开营救计划，并与搜救部门共享救援重要进展方面的信息。

(6) 在海难或空难中，联合救援协调中心是总的负责部门，执行并控制整个施救过程。联合救援协调中心可以为当地应急部门、救护部门、警察以及其他中立的资源和服务机构提供交接和联络的平台。

(7) 美加两国海岸警卫队一直都有合作，不管是在演习中，还是在实际事故处理中。这包括了海上资源和空中资源的共享，尤其是在两国的边界地带处理事故的配合与交流中的互相支持。

(8) 在事件中，协调员给海难现场指挥员下达指令，并且为各部门提供一个信息交流的平台。

（9）根据现场情况，海难现场指挥员有可能让下属去执行实际的船只操作工作，而自己更好地专注于事故现场管理。

（10）一旦训练有素的救援人员来到现场，海难现场指挥员或者运输管理员就要决定现场是否稳定，并将此信息传达给医疗类选指挥官以及可能接近现场或登上遇难船只的营救人员。

（11）如果说海难现场指挥员是联合救援协调中心的眼睛和耳朵，那么运输管理员就是现场指挥员的眼睛和耳朵。

（12）在抵达海难现场之前，就应该指派好医疗类选指挥官，而这位指挥官应该熟知海难现场管理和医疗类选的基本原理。

（13）在确定了现场的稳定性和风险之后，海岸警卫队的营救小组成员应该按要求协助运输管理员和医疗类选指挥官。

（14）根据训练有素的营救人员所采取的行动不同，这些旁观者可能会对营救的实施起到阻碍或帮助作用。现场营救人员应该派给旁观者他们力所能及的任务。

（15）灾难中未受伤的幸存者往往表现出惊人的恢复力，有时在救援过程中极能派上用场。

2. Translate the following passages into English

（1）A thousand tons of crude oil has spilled into the sea from an oil tanker，and the bad weather continues to complicate efforts to deal with the oil spilling.

（2）A stately liner can sail serenely through turmoil that would capsize even the sturdiest small vessel.

（3）The migrant vessels kept sinking. Last Friday，still another boat sank into the deceptively placid waters of the Mediterranean.

（4）A contingency plan is a plan devised for an outcome other than in the usual（expected）plan. It is often used for risk management when an exceptional risk that，though unlikely，would have catastrophic consequences.

（5）The United States Coast Guard（USCG）is a branch of the United States Armed Forces and one of the country's seven uniformed services. The Coast Guard has roles in maritime homeland security，maritime law enforcement，search and rescue，marine environmental protection，etc.

（6）When grounding，engine breakdown or fire is happening，the materials and reports below will be helpful for the ship owner to decide whether the professional salvors would be employed.

(7) Because of the nature of maritime travel, there is often a substantial loss of life in marine disaster. All ships, including those of the military, are vulnerable to problems from weather conditions, faulty design or human error.

(8) When working on lightning and rainy days, the operating personnel is likely to be hit by lightning stroke and unexpected casualty may be caused.

(9) A tidal wave caused by the earthquake hit the coast, causing catastrophic damage.

(10) After the triage area was set up, a command vehicle arrived with all the necessary supplies and equipment.

(11) Pollutants enter rivers and the sea directly from urban sewerage and industrial waste discharges, sometimes in the form of hazardous and toxic wastes.

(12) Oil spills may be owing to releases of crude oil from tankers, offshore platforms, drilling rigs and wells, as well as spills of refined petroleum products (such as gasoline, diesel) and their by-products or the spill of any oily refuse or waste oil.

(13) Ships can pollute waterways and oceans in many ways. Discharge of cargo residues from bulk carriers can pollute ports, waterways and oceans.

(14) According to relevant international practices, China should set up and continuously improve the early warning and contingency plan system of major marine disaster.

(15) They made the place a notorious center of piracy. The profits of piracy had gone but the piratical instinct remained.

Ⅴ. Exercise 2: Passages in Focus　练习 2：短文翻译

1. Translate the following passages into Chinese

七号灭火轮，作为一艘铝质船身的双体式救援船，于 1990 年交付使用。它是消防处唯一专供香港国际机场以外水域进行海上大型救援的船只，其主要职能如下：

• 在本港水域执行港口安全及救援任务，尤其是在发生大型海上事故（如船舶火灾或大型船只沉没等）时，把大批受灾人员/伤者从海上现场运送到陆上的安全地方或具备医疗设施的场所；

• 在发生海上火警时，支援其他灭火轮，提供救火服务；以及

• 在核动力船只访港期间，用作救援船。如发生紧急事故，该船会负责撤离核动力船只的船员，监察他们的辐射水平，以及在有需要的情况下，现场为船员提供简单的放射性洗消设施。

政府铝质船身船只的设计使用年限约为 15 年。七号灭火轮至今服役已逾 20 年。海事处在进行每年例行检修时发现，该船的船身和甲板已明显老化并出现锈蚀，其操作表现亦每况愈下，每年因机械故障而停航维修的时间亦由 2008 年的 24 天增加约 62%至 2011 年的 39 天。

为更有效地切合现今的运作要求，我们建议购置一艘具备下述更完备灭火与救援功能及装置的新船：

• 最高航速由现时每小时 27.5 海里提高至每小时 35 海里，以便更快抵达事故现场，尽快把受灾人员/伤者接送到陆上的安全地方或具备医疗设施的场所；

• 配备容量较大的救生筏（由目前可接载 320 人增至 420 人）用作大型救援用途，以配合 2 号邮轮码头的启用。邮轮一旦发生事故，可能会涉及超过一千名伤者；

• 配备一艘小艇，以便消防人员前往较浅水的事故现场评估情况及部署行动；

• 配备高效能空气过滤系统和辐射监察设备等装置，驾驶室/船舱亦会装置加压系统，以便在核动力船只访港期间做好准备，如有需要可用于救援行动。此外，新船亦会配备较完善的放射性消洗设施，以提升我们处理有关事故的能力，以及为前线人员提供更佳保障。

2. Translate the following passages into English

The definition of triage is: The sorting and allocation of treatment to patients, and especially battle and disaster victims, according to a system of priorities designed to maximize the number of survivors. If there are abundant rescuers and resources on scene then every casualty will get the same effort as would be given in a single casualty incident. Unfortunately in a disaster this is not the case. Consider that some people will survive no matter what care they receive, while some will die despite every effort. The key goal of triage is to identify those whose survival depends on early intervention and treatment. If we can do that and provide life saving interventions, we are maximizing the number of survivors.

Rescue units should pre-plan for high-risk or high impact incidents in their area of response. These could include major passenger or commercial traffic, coastal airports or other infrastructure with marine access or specific high risk activities. Consider the likely hazards, potential number of casualties, access and egress, and probable staging areas. Consider also the other agencies or resources that will respond, their responsibilities and capabilities and how you will work together.

The time to start thinking about triage is when first notification of the incident is received. Even on route, the Triage Officer must develop an idea of the overall seriousness of the situation. This is best done by considering the history of the event (e. g., speed of impact, severity of the fire, the length of immersion etc.), as well as the probable number of casualties. Based on this pre-arrival information rescuers may start to formulate a triage plan, however they must adjust this plan as required by what they find on arrival.

It has been said that incidents are won or lost in the first 10 minutes after arrival, when the most important decisions are made. After confirming scene stability, triage is the first task to be completed at the disaster scene. Because casualty condition can change, triage must be a continuous process, with casualties re-triaged throughout the course of the incident.

Unit Six　Marine Protection

第六单元　海洋保护

Ⅳ. Exercise 1: Sentences in Focus　练习 1：单句翻译

1. Translate the following sentences into Chinese

(1) 因为美国有成百上千的海洋保护区，而各联邦、各州、各部落、各地方都有自己的管理机构，所以各保护区内发生的事并不能互通有无，人们可能会错失合作的良机，也无法分享经验教训。

(2) 随着时间的推移，国家海洋保护区中心将继续与现有的美国海洋保护区合作，联结、加强、推进国家的海洋保护区项目。

(3) 诸如海洋自然保护区的某些区域则表现得更为苛刻，限制了捕鱼活动、贝类的采集活动，其他会把资源带离该区域的活动也受到了限制。

(4) 其特征包括保护重点、保护级别、永久性保护、恒常性保护和生态保护规模。

（5）此项网络工具能让你通过交互式地图查看海洋保护区清单上的地点位置及相关数据，或者通过详细的保护属性查询地点，或者分地区搜索查看位置。

（6）该制度要求商务部、内务部以及其他联邦机构联合各州、各领地、各部落、各渔业管理委员会、各海洋资源保护组织，以科学全面的方式共同发展代表美国海洋生态系统的国家海洋保护区体系。

（7）海洋保护区中心通过加强归化能力、管理能力、评估能力，将目标集中于推进海洋保护区财务管理。

（8）13个由美国国家海洋局直接管理的国家海洋保护区和28个由美国国家海洋局合作州管理的国家港湾研究保护区是受到保护的区域，也是娱乐、研究、教育的中心地带。

（9）这些目的受到了质疑，主要是因为管理保护区的花销和人类对海上货物和服务需求引发的保护活动之间的不协调。

（10）南太平洋海洋资源利用保护永久委员会旨在促进各成员之间学习和信息交流。

（11）强制将土著人口纳入海洋保护区的方法就是像澳大利亚使用的“土著保护区”。

（12）2009年4月美国建立了美国国家海洋保护区体系，该体系的目的是加强美国海洋资源、海岸资源和五大湖资源的保护。

（13）英国正大力发展英国海外领地的海洋保护区。

（14）该保护区将达到234 291平方千米，其中一半的区域将禁止捕捞活动。

（15）这两项关于3个地中海海洋保护区，相隔30年实施的评估体系，向世人说明了合理的保护是可以让具有商业价值但生长缓慢的红珊瑚在不足50米（160英尺）的浅水地带大量聚集的。

2. Translate the following sentences into English

（1）In the South Atlantic，deepwater marine protected areas were established to shield deep-water fish species and their habitats from fishing.

（2）States shall endeavor to participate actively in regional and global programs to acquire knowledge for the assessment of the nature and extent of pollution，exposure to it，and its pathways，risks and remedies.

（3）Despite national autonomy in the implementation of EU rules，however，it is ultimately often the EU bureaucracy and not the national governments that is held accountable for the ramifications of decisions taken at a national level.

(4) Coastal States may, in the exercise of their sovereignty within their territorial sea, adopt laws and regulations for the prevention, reduction and control of marine pollution from foreign vessels, including vessels exercising the right of innocent passage.

(5) The programme will also link with ongoing activities of the South Pacific Islands Applied Geoscience Commission with regard to data rescue and the development of a deep-sea minerals resource database and work on maritime boundaries (exclusive economic zones and extended continental shelf).

(6) No take zone placing restrictions on fisheries may be a way to enhance the productivity of remaining marine waters by increasing prey availability for dolphins, and minimizing disturbance to the seabed.

(7) At its meeting in February 2011, the Committee on Fisheries considered specific activities relevant to biodiversity conservation, including establishing marine protected areas and networks of areas, and carrying out impact assessments.

(8) A number of regional fisheries management organizations have adopted area closures and other area-based measures to address the impacts of fishing. The International Commission for the Conservation of Atlantic Tunas adopted several time/area closures, mainly to protect juveniles of tuna species such as bluefin, swordfish and bigeye.

(9) The United Nations Environment Programme, with partners, is executing and implementing the GEF Transboundary Waters Assessment Programme, which is aimed at developing assessment methodologies for the status and changing conditions in transboundary water systems.

(10) In the populous and economically prosperous coastal areas, China should create an integrated research and monitoring network of marine ecological environment comprised of environmental monitoring facilities, research institutions, laboratories, outdoor observatories, and ecological recovery demonstration projects.

(11) Assessments are important to better understand the status of, and trends in, the condition of marine ecosystems. In particular, assessments help gauge the vulnerability, resilience and adaptability of various ecosystems.

(12) The types of pollution include: pollution from land-based sources;

pollution from seabed activities; pollution from activities in the Area; pollution by dumping; pollution from vessels; and pollution from or through the atmosphere.

(13) The Intergovernmental Oceanographic Commission has organized training courses on data and information management, sea-level data analysis, modelling, marine biodiversity and the application of remote sensing to coastal management.

(14) With respect to biodiversity in the deep and open oceans, the International Union for Conservation of Nature participates in the Global Ocean Biodiversity Initiative.

(15) The Global Ocean Observing System is a permanent global system for observations, modelling and analysis of marine and ocean variables to support operational ocean services worldwide.

Ⅴ. Exercise 2: Passages in Focus 练习 2：短文翻译

1. Translate the following passages into Chinese

现有资料表明，在过去十年中，保护区覆盖的面积大量增加。然而，许多生态区，特别是海洋生态系统中的生态区，仍未能得到充分保护，保护区的管理效力仍然不稳定。在232个海洋生态区域中，18%的达到保护区覆盖面积至少有10%的目标，而有一半的覆盖率还不到1%。海洋保护区的总数大约已达到5 880个，面积470万平方千米，占世界海洋面积的1.31%。全球海洋保护区主要由数目不多的面积非常大的海洋保护区组成，几乎所有海洋保护区都在国家管辖范围内。

最近的一份报告强调了海洋保护区的成本和收益。实施、建立、维护和适应性管理海洋保护区的成本可能很高，但有关设立和管理海洋保护区和保护区网络成本的数据仍然有限。2002年，每年管理单一海洋保护区的费用估计为9 000美元至600万美元。2004年，达到覆盖保护区面积20%到30%目标的全球海洋保护网络的费用估计为50亿美元至190亿美元。就收益而言，报告列举了海洋保护区在渔业、旅游、精神、文化、历史、美学价值、减灾、科研、教育以及海洋意识与保护管理工作等方面的收益。

2010年9月，《保护东北大西洋海洋环境公约》缔约方商定，指定6个公海海洋保护区：米尔恩海隆复合区、查理-吉布斯南部断裂带、阿尔泰公海、安蒂阿尔泰公海、约瑟芬公海以及亚速尔群岛公海以北大西洋中脊，自2011年4月12日起生效。这些海洋保护区连同国家管辖范围内海洋保护区网络，占《保护东北大西洋海洋环境公约》总覆盖面积的 3.1%。这些海洋保护区中有一些是位于沿海国家外部的大陆架上的。查理-吉布斯南部断裂带和米尔恩海隆复合区旨在保护和养护海床和上方水域的生物多样性和生态系统，其他四个海洋保护区的建立是为了保护和养护有关海域上方水域的生物多样性和生态系统，与葡萄牙采取的海底防护措施互为协调和补充。

2. Translate the following passages into English

Marine protected areas and area networks, as part of broader coastal and ocean management frameworks, are considered a key tool to help ecosystems remain healthy and perform ecological functions by protecting critical habitats. However, for marine protected areas to achieve their objectives, they need to be designed and managed effectively, taking into consideration the socio-economic needs of stakeholders. They also need to be part of an effective broader framework that addresses management across all sectors, and to act in synergy with other tools.

At its tenth meeting, the Conference of the Parties to the Convention on Biological Diversity adopted a new strategic plan to achieve a significant reduction of biodiversity loss by 2020. Several of the 20 targets of the plan are relevant to marine biodiversity, including in areas beyond national jurisdiction. In particular, it was agreed that, by 2020, at least 10 percent of coastal and marine areas, especially areas of particular importance for biodiversity and ecosystem services, would be conserved through effectively and equitably managed, ecologically representative and well-connected systems of protected areas and other effective area-based conservation measures, and integrated into the wider seascapes.

The IOC secretariat, in its contribution, stated that a network of marine protected areas beyond areas of national jurisdiction or any other management action in such areas would require a monitoring system and a strong evidence base for policy-setting. Frequent and reliable observations are essential, as oceanographic features are dynamic. In that regard, fixed-boundary marine protected areas would not give the protection necessary to preserve pelagic biodiversity, and a solution being explored by IOC was therefore the use of dynamic marine protected area boundaries, following the example of electronic nautical charts. It also noted that the enforcement of marine protected areas beyond areas of national jurisdiction depended on the availability of vessel-tracking systems and remote sensing tools. IOC, with the Marine Board of the European Science Foundation, has established a working group to provide a framework to inform, engage and empower stakeholders in future marine protected area planning. The working group is reviewing and synthesizing the factors that should be considered for placing and establishing marine protected areas; reviewing criteria for the assessment of established areas; and developing a checklist of criteria for evaluating the efficacy and performance of an area.

Unit Seven Island Tourism

第七单元 海岛旅游

Ⅳ. Exercise 1: Sentences in Focus 练习1：单句翻译

1. Translate the following sentences into Chinese

(1) 岛屿的物理和气候特征对越来越多的游客产生了独特的吸引力。

(2) 资源缺乏多样性，无法吸引广泛的国际游客，结果导致了岛上游客的季节性流动。

(3) 但是对于和旅游业有关的发展政策来说，岛屿吸引游客的因素也是其长期成功的挑战。

(4) 对于当地居民而言，能够理解游客的需求，吸引大量游客的到来十分重要，同时，他们也要了解旅游业对岛屿所造成的影响。

(5) 气候和地理上的隔绝状态对于岛屿的经济发展不利。

(6) 不容易到达是岛屿的另外一个特征，可能会造成交通费用高，游客减少，供应缺乏，物价高和公共服务方面的问题。

(7) 实际上，由于生态系统和自然美景是岛屿的主要优势所在，要想维持在旅游产业的竞争力，它们就要对其进行维护。

(8) 岛屿通常具有边缘性、孤立性、脆弱性、资源短缺、劳动力有限和运输代价高等特征，所有这些都是竞争上的劣势。

(9) 旅游业的很多学者认为旅游业已经成为帮助岛屿克服面积约束的重要途径。

(10) 旅游业在理论上提供了一种灵活的、可以不断再生的资源，可能会解决这些问题中的一部分。

(11) 旅游业不是解决岛屿所有问题的灵丹妙药，除了积极影响之外，开发旅游业还会有一些负面的影响。

（12）因此，在积极的方面，旅游业可以提高人们的环境意识，激励环保政策的出台。

（13）季节性的波动会造成季节性的就业、设施利用率不高和过度拥挤，并给交通系统和公共服务带来压力。

（14）当地人开始模仿游客的文化，贬低他们自己的文化和遗产的价值和历史，对于这一现象，人们有清醒的认识，将其看作是一种文化上的影响。

（15）由于到岛上旅游的决定不受岛屿自身的支配，世界市场条件的轻微波动常会影响到岛屿。

2. Translate the following sentences into English

(1) The advent of industrial fishing threatens many marine species with extinction.

(2) Many islands in the South Pacific hold irresistible allure to honeymooners.

(3) Irresponsible fishing practices are the major forces endangering the sustainability of the fisheries.

(4) As the government looks to cut down emission levels in the country, car makers are gearing up to cater to growing demand for environmentally-friendly cars.

(5) The importance of this park derives from its wealth of flora and fauna.

(6) A Wall Street Journal editorial encapsulated the views of many Conservatives.

(7) The incentive system rewards individuals based on performance and achievement of individual goals.

(8) Many of the springs have curative properties, one of them discharging about 3,000 gallons of water per minute.

(9) The hydrosphere is, in fact, continuous, and the land is all in insular masses: the largest is the Old World of Europe, Asia and Africa; the next in size, America; the third, possibly, Antarctica; the fourth, Australia.

(10) Companies are increasingly keen to contract out peripheral activities like training.

(11) In the nature reserve for white whales one can swim with these wonderful creatures or even touch them.

(12) The revenue from tourism is biggest single item in the country's invisible earnings.

(13) The first step to revitalize tourism is to attach importance to the protection of the ecological environment.

(14) A stagnant economy combining with a surge in the number of teenagers is likely to have contributed to rising crime levels in the US.

(15) Nature has an amazing power to calm even the most unsettled mind down and set anyone in the mood of serenity.

Ⅴ. Exercise 2：Passages in Focus 练习2：短文翻译

1. Translate the following passage into Chinese

现在，加勒比群岛的经济依赖于旅游业，旅游业被称为其“增长引擎”。对所有的加勒比国家来说，旅游业都做出了巨大的贡献，而对于其中很多国家来说，旅游业是国家收入的最大贡献者，如安提瓜和巴布达、巴哈马和维尔京群岛。旅游业是稳定的收入来源，只有在发生飓风或者西方世界经济衰退时才会暂时中断。旅游业为当地的农业、渔业和零售业提供支撑。例如，巴巴多斯已经从农业为主的经济转向支撑旅游业的服务经济。截止到2006年，旅游业对巴巴多斯经济的贡献是蔗糖业的10倍，两者分别是1.67亿美元和1 450万美元。在就业方面，这个地区11.3%的工作直接或间接依赖于旅游业，常被称为“世界上最依赖于旅游业的地区”。

这一地区的旅游项目通常和海洋热带气候有关，如在珊瑚礁进行轻便潜水和浮潜、乘船游览、帆船和海钓。陆上项目有高尔夫、植物园、公园、石灰岩洞、自然保护区、徒步旅行、自行车和骑马。文化活动包括狂欢节、钢鼓乐队、雷鬼音乐和板球。由于这些岛屿比较分散，直升机或飞机旅游很受欢迎。这一热带地区所特有的旅游项目包括历史性的殖民地种植园房屋、制糖厂和朗姆酒酿造厂。加勒比的烹饪是多种烹饪风格的融合，炖羊肉是几个岛屿的招牌菜。

很多游客来岛上度蜜月或者是旅行结婚。早期的海滨度假区主要是用来治病健身，洗海水浴，呼吸温暖的新鲜空气。在旅行指南上，巴巴多斯被称为“西印度群岛的疗养院”，因为这里水质清冽、海风怡人，没有瘴气。

20世纪20年代，游客到加勒比群岛度假是为了享受日光浴。当时日晒被认为是健康的，在富人中间，被晒成棕褐色是性感的象征。在二战之前，每年有十万多游客来到这一地区。

随着自由贸易政策的出台，加勒比地区的香蕉、蔗糖和矾土失去了价格上的优势，旅游业在经济上成为重要产业。在联合国和世界银行的鼓励之下，从20世纪50年代开始，加勒比地区的很多政府大力发展旅游业，推动其发展中国家经济的发展。1951年成立了加勒比旅游协会。税收上的优惠政策鼓励外国人投资建设酒店和基础设施，由新成立的旅游部门予以支持。

2. Translate the following passage into English

As a member of the Alliance of Small Islands States, the Maldives Islands is one of the countries most vulnerable to the consequences of climate changes. This fact is due to the geographical and economic characteristics that the Maldives share with the other members of the AOSIS. They are small portions of land, surrounded by ocean and frequented located in regions that are prone to natural disasters. Besides that, they have poorly developed infrastructures, limited resources and are economically dependent to other countries.

In the case of the Maldives Islands the economy and wellbeing of the population is highly dependent on the good conditions of the environment and any change on climate conditions can be dangerous. The tourism is responsible directly and indirectly for a high portion of government revenues. Furthermore, 20% of the population depends on fisheries as their major income earning activity, and all the infrastructure of the islands is at a low elevation of 1.5 m above the sea level.

Therefore, any change on the patterns of environment can be seen as a threat to the wellbeing of the Maldivian population. Examples of this are the flood of 1987, which caused damages of US $4.5 million to the Male' International Airport and the Tsunami in 2004, which affected the tourism sector in US $230 million. Apart from the structural damage, there are several damages that threaten the life of the Maldivian population. The changes on precipitation patterns and flooding caused an increase on diseases, like dengue and others. Due to changes on temperature and precipitation the Maldivian population now experiences food shortages each year, lasting longer than 10 days.

Thus, what we can see is that even though the Maldives Islands, together with the other small islands states, is one of the countries contributes the least to the green house gas emissions, with less than 1% of the global emissions, it is one of the most vulnerable to the effects. Instead of applying its resources to alleviate poverty and improve life conditions, the government is now obliged to use it in projects in order to deal with the climate change consequences.

The discussion about the sea level rise is intense among the scholars. The different ways to analyze the situation lead to a lack of consensus about the relationship between climate change and sea level rise. However, the Maldives Islands do face an enhanced threat of inundation from a mean sea level rise in this century, exacerbating the existing risks from a range of transient sea level changes in combination with the wave and swell which are the primary causes of erosion.

Unit Eight Marine Laws and Regulations

第八单元 海洋法律法规

Ⅳ. Exercise 1: Sentences in Focus 练习 1：单句翻译

1. Translate the following sentences into Chinese

（1）在解决公海海底的资源开采、航行权、经济管辖权或其他任何亟待解决的问题之前，《海洋法公约》必须先面对一个主要问题，即界线的划定。

（2）传统上，小国和那些没有大规模远洋海军或商船队的国家希望能够扩大领海范围，这样就可以保护它们的沿海水域不受那些拥有大规模远洋海军和商船队的国家的侵犯。

（3）随着大会的进展，12 海里领海的提议被越来越多的国家所接受，并最终为所有的国家所接受。

（4）《海洋法公约》规定军舰和商船有权“无害通过”沿海国家的领海。

（5）除了在领海范围之内享有执法权以外，沿海国家还有权在领海之外从海岸算起 24 海里范围内的区域行使某些权利，目的是为了防止某些违规行为，履行治安权。

（6）航行权的问题被认为可能是最关键的问题，也是让参与制定《海洋法公约》的谈判者最感到为难的问题。

（7）国家通常会将海岸之外的部分海域视为其领土的一部分，作为保护区域进行巡逻，驱逐走私者、军舰和其他入侵者。

（8）到了 20 世纪 60 年代末，12 海里领海的趋势已经在世界范围内逐渐出现，大多数国家宣称在此范围内拥有主权。

（9）在第三次联合国海洋法会议上，海峡通行权的问题让主要的海军强国站到了一边，控制着狭窄海峡的沿海国家站到了另一边。

（10）《海洋法公约》中形成的妥协是一个新概念，结合了在法律上已经被接受的在领海无害通行和公海航行自由的条款。

（11）对于从海岸算起200海里之内水域、洋底和底土中的所有资源，沿海国家有权开采、管理和保护。

（12）浮游植物是鱼类的主食。在靠近陆地的地方形成最强的洋流，以及强大的离岸风引起的冷水上涌，都会把深层的浮游植物带到上层。

（13）沿海国家希望控制毗连水域的渔业捕捞，这是建立专属经济区的主要驱动力。

（14）随着长期利用的渔场开始出现耗竭的迹象，随着远洋渔船来到被当地渔民视为传统渔场的水域以及随着竞争的加剧，冲突也开始增多。

（15）每个沿海国家要确定其专属经济区范围内每种鱼类的总可捕量，还要评估其捕捞能力以及自己能够捕捞什么，不能捕捞什么。

2. Translate the following sentences into English

（1）Like land plants，phytoplankton have chlorophyll to capture sunlight，and they use photosynthesis to turn it into chemical energy.

（2）Only when environmental deterioration or resource depletion translate into economic decline do we seem to notice the problem.

（3）These chemicals have a detrimental impact on the environment.

（4）At that time the island was under foreign dominion.

（5）United Nations spokesman expressed concern and voiced hope for an expeditious resolution of the matter.

（6）China owns indisputable sovereignty over the Nansha Islands and their adjacent waters.

（7）It is known to all that these islands have always been under Chinese jurisdiction.

（8）Most navigable waterways require dredging because of siltation. This is true when navigation channels penetrate through the littoral zone.

（9）Many fishermen abandoned fishing for more lucrative employment in the booming construction industry.

（10）Fishery activities should not infringe upon the fishing rights of other countries.

（11）The brightly-coloured boats ply between the islands.

（12）These countries' geographical proximity to China is a great advantage for us in developing business relations with them.

(13) River habitats，called riparian systems，often contain lush vegetation and a wealth of animal species.

(14) During their talks，the two presidents discussed the transit of goods between the two countries.

(15) The country's unilateral decisions to increase their mackerel catches were severely criticized by its neighbors.

Ⅴ. Exercise 2：Passages in Focus 练习 2：短文翻译

1. Translate the following passages into Chinese

当沿海国主管部门有充分理由认为某外国船舶违反该国的法律和规章时，可对该船进行紧追。此项追逐须在该外国船舶或其小艇之一在追逐国的内水、领海或毗连区内时开始，而且只有追逐未曾中断，才可在领海或毗连区外继续进行。当外国船舶在领海或毗连区内接获停驶命令时，发出命令的船舶并无必要也在该领海或毗连区内。如果外国船舶是在《领海与毗连区公约》第 24 条所定义的毗连区内，追逐只有在设立该区所保护的权利遭到侵犯的情况下才可进行。

紧追权在被追逐的船舶进入其本国领海或第三国领海时立即终止。除非追逐的船舶以可用的实际方法认定，被追逐的船舶或其小艇之一或作为一队进行活动而以被追逐的船舶为母船的其他船艇是在领海界限内，或者根据情况在毗连区内，否则紧追不被认为已经开始。只有在外国船舶可视听的距离内给出视听停驶信号后，紧追方可开始。

紧追权只可由军舰、军用飞机或为此目的而经专门授权的其他政府船舶或飞机行使。在飞机进行紧追时：

(a) 应比照适用本条第 1 款至第 3 款的规定；

(b) 发出停驶命令的飞机，除非其本身能逮捕该船舶，否则须积极追逐船舶直到其所召唤的沿海国船舶或另一飞机前来接替追逐为止。当飞机只发现船舶犯法或有犯法嫌疑，如果该飞机本身或由其他飞机或船舶接着无间断地进行追逐时，未命令该船停驶和进行追逐，则不足以构成在公海上逮捕的理由。

在一国管辖范围内被逮捕并被押解到该国某一港口以便主管当局审讯的船舶，不得仅以在航行中由于情况需要而曾被押解通过公海的一部分为理由而要求释放。

在无正当理由行使紧追权的情况下，在公海上被命令停驶或被逮捕的船舶，对于可能因此遭受的任何损失或损害应获得赔偿。

2. Translate the following passages into English

Article 1 This Law is enacted for the People's Republic of China to exercise its sovereignty over its territorial sea and the control over its contiguous zone, and to safeguard its national security and its maritime rights and interests.

Article 2 The territorial sea of the People's Republic of China is the sea belt adjacent to the land territory and the internal waters of the People's Republic of China.

The land territory of the People's Republic of China includes the mainland of the People's Republic of China and its coastal islands; Taiwan and all islands appertaining thereto including the Diaoyu Islands; the Penghu Islands; the Dongsha Islands; the Xisha Islands; the Zhongsha Islands and the Nansha Islands; as well as all the other islands belonging to the People's Republic of China.

The waters on the landward side of the baselines of the territorial sea of the People's Republic of China constitute the internal waters of the People's Republic of China.

Article 3 The breadth of the territorial sea of the People's Republic of China is twelve nautical miles, measured from the baselines of the territorial sea.

The method of straight baselines composed of all the straight lines joining the adjacent base points shall be employed in drawing the baselines of the territorial sea of the People's Republic of China.

The outer limit of the territorial sea of the People's Republic of China is the line every point of which is at a distance equal to twelve nautical miles from the nearest point of the baseline of the territorial sea.

Article 4 The contiguous zone of the People's Republic of China is the sea belt adjacent to and beyond the territorial sea. The breadth of the contiguous zone is twelve nautical miles.

The outer limit of the contiguous zone of the People's Republic of China is the line every point of which is at a distance equal to twenty-four nautical miles from the nearest point of the baseline of the territorial sea.

Article 5 The sovereignty of the People's Republic of China over its territorial sea extends to the air space over the territorial sea as well as to the bed and subsoil of the territorial sea.

Article 6 Foreign ships for non-military purposes shall enjoy the right of innocent passage through the territorial sea of the People's Republic of China in accordance with the law.

Foreign ships for military purposes shall be subject to approval by the Government of the People's Republic of China for entering the territorial sea of the People's Republic of China.

Article 7 Foreign submarines and other underwater vehicles, when passing through the territorial sea of the People's Republic of China, shall navigate on the surface and show their flag.

Article 8 Foreign ships passing through the territorial sea of the People's Republic of China must comply with the laws and regulations of the People's Republic of China and shall not be prejudicial to the peace, security and good order of the People's Republic of China.

Foreign nuclear-powered ships and ships carrying nuclear, noxious or other dangerous substances, when passing through the territorial sea of the People's Republic of China, must carry relevant documents and take special precautionary measures.

The Government of the People's Republic of China has the right to take all necessary measures to prevent and stop non-innocent passage through its territorial sea.

Cases of foreign ships violating the laws or regulations of the People's Republic of China shall be handled by the relevant organs of the People's Republic of China in accordance with the law.

Article 9 The Government of the People's Republic of China may, for maintaining the safety of navigation or for other special needs, request foreign ships passing through the territorial sea of the People's Republic of China to use the designated sea lanes or to navigate according to the prescribed traffic separation schemes. The specific regulations to this effect shall be promulgated by the Government of the People's Republic of China or its competent authorities concerned.

Article 10 In the cases of violation of the laws or regulations of the People's Republic of China by a foreign ship for military purposes or a foreign government ship for non-commercial purposes when passing through the territorial sea of the People's Republic of China, the competent authorities of the People's Republic of China shall have the right to order it to leave the territorial sea immediately and the flag State shall bear international responsibility for any loss or damage thus caused.

Unit Nine　Marine Engineering

第九单元　海洋工程

Ⅳ. Exercise 1：Sentences in Focus　练习 1：单句翻译

1. Translate the following sentences into Chinese

(1) 那么，怎样才能获取并利用这一丰富的能源呢？60 多年来，人们对利用这一能源的几种不同的方式进行过研究并加以实施。随着燃料化石的日益枯竭以及价格持续上涨，这些能源在成本上十分具有竞争力——这种“燃料”无需成本且非常清洁。

(2) 传统上，在所有可利用能源中，潮汐能利用成本相对要高，而且，潮差和流速达到足够高度的选址十分有限，潮汐能的总体利用率因而受到限制。

(3) 然而，在设计和涡轮机技术方面，许多新近的技术上的发展和进步表明，潮汐能的总利用率可能比之前的预期要高得多，经济与环境成本也可能会降到具有竞争力的水平。

(4) 潮磨包括一个蓄水池，在涨潮时，潮水通过水闸涌入而蓄满，在退潮期，潮水经过水轮溢出而流空。

(5) 潮汐能的缺陷在于其容量小，由于潮汐周期为 12.5 小时，无法满足峰值期需求。

(6) 水坝包括一个水闸，开启后让潮水流进水池，然后再将水闸合上。随着潮位降低，水头（水池中的高位水）利用传统的水力技术，驱动涡轮机发电。

(7) 潮差可能在一个较大的范围内（4.5～12.4 米）因位置不同而发生变化。对涡轮机来说，要保证至少 7 米的潮差，才能做到运行经济，水头足够。

(8) 在 20 世纪 60 年代末，他们的核能计划项目获得巨大发展之前，拉郎斯应该是法国众多潮汐发电厂中的一个。

(9) 潮汐能属可再生电力来源，不会导致引起全球气候变暖的气体的排放，也不会造成与发电的化石燃料相关的酸雨。

(10) 然而，因海湾或河口筑坝拦水而引起的潮汐流的变化可能会对水生和海岸线生态系统，以及航行和个人休闲活动带来不良影响。

（11）迄今为止，为确认潮汐能计划是否会对环境造成影响的少数几项研究都已断定每一特定的位置不同，其影响在很大程度上取决于当地的地形地貌。

（12）决定潮汐能位置成本效益的主要因素是所需拦河坝的规模（长度和高度），以及潮水高低潮间的高度差。这些因素都可以用一种被称为位置上的“吉不列比”来表述。

（13）布局离岸蓄水，而非采用传统“拦河坝”方式，消除了阻碍商业规模的潮汐能发电厂兴建的环境与经济问题

（14）每天都会出现两个高潮和两个低潮，因而，潮汐能电厂发电的特点表现为每十二小时出现一个最大发电量周期，其间，第六个小时是不发电的。

（15）由于月运周期和引力的影响，潮汐流尽管是一个可变量，但十分可靠，并具可预测性。其能源对具有多种来源的电力系统可以做出非常有价值的贡献。

2. Translate the following sentences into English

(1) NPV and PI should always be stated with their respective discount rate.

(2) Zhang Zhenyu, spokesman of Shanghai Flood Control Headquarters, said that the city is considering building a floodgate near the Yangtze River estuary.

(3) Global warming will cause dramatic climate changes including more frequent floods, heat waves, and droughts.

(4) Turkey plans to harness the waters of the Tigris and Euphrates rivers for big hydro-electric power projects.

(5) The reservoir impoundment will change the geologic environment and bring about environmental problems.

(6) Beneath the heavenly sphere this solar cycle is mirrored in the lunar cycle of wax and wane and tidal ebb and flood.

(7) The models are based on weather, ocean current and spill data from the U. S. Navy and the National Oceanic and Atmospheric Administration.

(8) The Ocean Thermal Energy Conversion (OTEC) process utilizes temperature differences in large bodies of water to spin turbines, creating a steady source of renewable energy.

(9) Though construction and maintenance costs tend to be higher for

offshore turbines than land-based ones, offshore wind farms can benefit from lower transmissions costs since many major cities sit on the coast.

(10) Wind power has emerged as a viable renewable energy source in recent years—one that proponents believe could lessen the threat of global warming.

(11) These sluice gates have never been tested on a dam of this magnitude. Nowhere in the world is there a successful model that deals with sediment on this scale.

(12) Ocean energy is a new clean energy resource which is environmentally-friendly and renewable. It includes tidal energy, ocean current energy, wave energy, sea salt and sea water temperature.

(13) So as we attempt to jump-start the economy of 2009, we should recognize both the risks and advantages inherent in a robust credit industry.

(14) In recent years, with widespread of new type pipe material and decoration of buildings, there is insufficient pressure head of water supplied at top floors in high-rise buildings.

(15) It plays an increasingly important part in meteorology for cloud, precipitation, hail and thunderstorm detection as well as the navigation of aircraft and ships.

Ⅴ. Exercise 2: Passages in Focus　练习 2：短文翻译

1. Translate the following passages into Chinese

2014 年，芬迪海洋能源研究中心实现了一个重大转折：安装了水下电力电缆，沿米娜通道海底铺设了四根电缆，从而使该中心具备了世界上最大的潮汐能输电能力。四根 34.5 千伏的电缆，总长度为 11 千米，其总容量达 64 兆瓦，相当于高峰潮汐流时两万个家庭的用电需求。这种水下基础设施将使小型涡轮机阵列能连接到电网。

此外，芬迪海洋能源研究中心所有四个开发商的发电总量为 17.5 兆瓦（米娜能源：4 兆瓦；黑岩潮汐能公司：5 兆瓦；加拿大亚特兰蒂斯运营公司：4.5 兆瓦；开普夏普潮汐能合资企业：4 兆瓦）的开发嵌入电价项目于 2014 年获得批准。项目的批准使得所有开发商与诺瓦史寇夏电力公司签订了一个为期 15 年的电能购买协议。开放水力公司和黑岩潮汐能公司已经在诺瓦史寇夏设立了办事处，并开始招聘员工。开放水力公司的组件采购与装配正在有条不紊地进行。

芬迪海洋能源研究中心已经完成建造迷你着陆器——为芬迪先进传感器技术（FAST）平台计划而独创的第一个水下平台，它设计用来通过电缆连接对海洋环境进行实时测量。芬迪先进传感器技术平台会对新兴全球需求做出回应以识别和确认合适的潮汐能位置以及监测现有位置的环境状况，这些都需要使用一定的技术和方法以在高流量潮流的极端条件下采集数据。

芬迪海洋能源研究中心继续实施环境影响监测计划，并完成了第二份涵盖2011—2013期间的研究报告。如果在流潮汐技术发展成一个较大的并具有商业规模的项目，开发就会顺利展开。由于米娜通道的涡轮发动机已经安装到位，芬迪海洋能源研究中心正竭尽全力地了解会对环境产生何种影响以及公布这些对大众的影响。

2. Translate the following passages into English

Most people don't realize that the ocean is a vast reservoir of energy that might be put to human use. Harnessing the sea's energy is the objective of several bold concepts that may help meet the energy needs of the twenty-first century.

Mill wheels were used since ancient times to harness tidal energy, the tremendous energy contained in the normal ebb and flow of the tides. Modern schemes call for the construction of large barriers across narrow bays and river estuaries in areas where the tidal range is high, at least 3 m. Water moving in with the high tide is caught behind the barrier, and locks are opened to release the water at low tide. The flowing water drives turbines that generate electricity, just as in the hydroelectric plants in river dams. The mechanical energy contained in the tide is thus used to obtain electricity. One such electrical generation plant has been operational in Brittany, northwestern France, since 1966. A few other facilities have been built on an experimental basis. Large projects are envisioned on the River Severn in western England, the Bay of Fundy in eastern Canada, and other suitable areas.

The use of tidal energy is pollution-free and relatively efficient, but the resulting changes in the tidal patterns can be highly destructive to the nearby environment. The rich marshes and mudflats in estuaries may be damaged or destroyed. Pollutants from other sources tend to accumulate upstream because normal tidal flushing is restricted. River flows can also be altered, increasing the risk of floods inland. On the other hand, the artificial lagoon created upstream can be used in recreation, as in the case of the French plant.

Unit Ten　Maritime Transport

第十单元　海洋运输

Ⅳ. Exercise 1: Sentences in Focus　练习 1: 单句翻译

1. Translate the following sentences into Chinese

(1) 因此，无论是在历史上还是在当代，进行商品贸易始终都是海上交通网络构建的主要动力。

(2) 随着十九世纪中期蒸汽机的广泛应用，船舶航行不再受主导风向的制约，海上贸易网亦随之显著扩张。

(3) 与陆空运输模式类似，海洋运输也有自己的运行空间。这种空间既指其物理意义上的地理空间，也指其市场控制的战略空间和商业运作空间。

(4) 因为洲际间的海上运输总是尽量按照大圆弧长的路线航行，其结果是，海上航线在地球水体表面绘出了许多条弧线。

(5) 影响水运的最新技术变革重点放在了改造水上通道（如疏浚港口通道，增加其深度），提升船舶（如集装箱船、油轮、散货船）的吨位、自动化和专业化程度，以及发展大型港口码头的设施以支持海上运输的技术要求上。

(6) 港口与河流交通的关联性不大，但河流枢纽中心正在与海陆运输，特别是集装箱运输不断整合。

(7) 尽管在选定的河流主干道（如长江）上有定期的航运服务，但其水路客运潜力仅局限于河流旅游（如坐船游览）。

(8) 然而，内陆区域未必被排除在国际海运贸易之外，但是持续增高的运输成本可能有碍其经济发展。

(9) 此外，内陆的概念有时是相对的，因为如果一个沿海国家的港口基础设施不足以应付其海上贸易，或者它的进出口商使用的是第三国的港口，那么这个沿海国家就可被视为是相对意义上的内陆国。

(10) 钟摆式海运服务涉及至少两个海域范围内的一系列连续的沿途到港停靠，通常包含跨洋的海运服务，以一个连续循环的营运模式构成。

(11) 在集装箱运输之前，装卸船舶是一项很昂贵和耗时的工作，通常货船在码头停靠的时间比其在海上航行的时间还长。

(12) 最近的另一个趋势是，随着支线船舶在各主要海上中间枢纽聚集，多条航线已经一体化和专业化了。考虑到偏离主要海运航线给沿途停靠的服务长度和频率带来的负面影响，这种趋势在欧洲（地中海、北海和巴尔地海）表现得尤为明显。

(13) 在洲际航空运输时代之前，跨陆客运服务是由班轮客运船舶承担的，主要是横跨北大西洋。

(14) 海上交通运输一般以“载重吨”来计量。“载重吨”指的是装载在“空”船上的货物的重量，它不能超过该货船营运设计的限制。这个限制常以一条载重线来标识，也是船的最大吃水线；它没有算入船的自重，但包含了燃料和压载水的重量。

(15) 传统意义上的散货运输有着单一的货源、目的地和客户，易于形成规模经济。除了能源贸易和作为垂直一体化生产过程的一部分之外（如从油田到港口再到炼油厂的运输），散货运输服务往往都是不定期的。

2. Translate the following sentences into English

(1) In the hub-and-spoke transportation model, it is small and medium sized container ships that finish the feeder line tasks; these ships can fulfill the cargo transportation tasks between hub ports and hinterland, hub ports and feeder ports well.

(2) Crews are waiting for the transshipment containers from the feeder barge being loaded to this voyage.

(3) The aim of the optimization design is to solve the contradiction between inner storage space of ship body and structure deadweight and maneuverability of vessel.

(4) This article presents a method of calculating the standard deviation, qualified rate and accuracy of the automatic weather station (AWS) measurement results.

(5) There also exist disadvantages in pendulum marine service, such as the risk of empty trips and extended service time between distant ports along the route.

(6) These developments result in the need for a wireless backhaul transport network system to facilitate interoperability, end-to-end security and end-to-end quality of service.

(7) The overflow dredging method is predominantly used to dredge silt in the channel and, generally, only with outgoing tide.

(8) On average, 40 U. S. navy vessels make port call in Hong Kong each year, and the sailors spend millions of dollars in stores, hotels, restaurants and bars.

(9) One of the Shenzhen market's competitive advantages in the past was its proximity to Hong Kong, a source of thoughts, ideas, advice and investment.

(10) With the development of trade and the improvement of the containerization of a piece of groceries and some bulk cargo, the situation of lasting growth has appeared all the time in the quantity of goods transportation of the container.

(11) Object oriented technology should be adopted in system program design because it can both decrease the number of code and make all system credible, effective, all structure clear, prone to maintenance and extension.

(12) We will take practical steps to seek steady development in accordance with the principles of reciprocity and efficiency.

(13) Long-distance express freight trains are equipped with facilities for rapid loading and unloading of goods; they are an integral part of the entire transportation system, undertaking the task of cargo transportation between the industrial centers and the seaports.

(14) An infrastructure that is optimized for database cloud delivery emphasizes simplicity and efficiency through automation and hardware standardization.

(15) Many transport workers and city governments begin to give priority to the public transit development, to achieve the sustainable development and urban land use intensification, and to avoid the heavy car trips in western developed countries.

Ⅴ. Exercise 2: Passages in Focus　练习 2：短文翻译

1. Translate the following passages into Chinese

全球海运业由约十万艘百吨以上的商业船舶提供服务。这些商船分为以下四大类：

客运船舶在历史上发挥了重要的作用，因为它们是当时可用于长途运输的唯一方式。在当代，客船可分为两类：客运渡轮以穿梭式服务为人们提供相对短程的水域内（如河流或海峡）的交通；游船则搭载乘客旅行度假，持续时间可长可短，通常都在数天以上。前者往往是更小、更快的船只，而后者通常是非常大容量的船舶，且拥有全方位的设施。2012 年，约有 2 030 万名乘客接受了游船服务，说明这是一个有很大增长潜力的行业，因为它服务于多个季节性市场，船队全年都有工作部署。

散装货船是专为运载特定商品而设计的船只，可分为液体散装货船和干货散装货船。它们包括了海上最大的船舶。如最大的油轮，即超大型油轮（ULCC）的载重吨（dwt）高达 50 万吨，更典型的吨位则在 25 万到 35 万载重吨之间；最大的干散货船的载重吨为 40 万吨左右，更典型的吨位也在 10 万和 15 万载重吨之间。液化天然气（LNG）技术的出现使得启用专用船舶进行天然气海上贸易成为可能。

一般货船是专为运载非散装货物而设计的船只。因为装载和卸载速度极慢，所以传统的一般货船小于 1 万载重吨。自 20 世纪 60 年代以来，这些船只已被集装箱船所取代，因为集装箱船可以更迅速和更有效地装载，因而可以更好地应用规模经济原则。跟任何其他类型船舶一样，大型集装箱船需要更大的吃水量，而目前最大的一般货船需要的吃水量为 15.5 米。

滚装船是为了便于直接装载汽车、卡车和火车而设计的船只。最初滚装船以渡船形式出现，它们用于深海贸易，比典型的渡轮大得多。最大的滚装船是汽车滚装船，用于将车辆从装配厂运输到各主要市场。滚装船的容量是以它可为运载的车辆提供的停车位的数量来衡量的，多以车道的米数来计量。

船舶类型的区别还可根据其被配置的服务类型进一步分类。散装船往往在两个港口之间按照固定的时间表运行，或者按航次运行，以反映需求的波动。这种需求可能是季节性的，如粮食运输，也可能是商机，如工程货物运输（例如运载建筑材料）。一般货船则提供定期班轮运输服务（即船舶在固定停靠的港口之间按常规时间表运营），或作为不定期货船运营（即船舶没有固定的运行时间表，而是根据货物是否到位来进行港口间的运输。）

2. Translate the following passages into English

Maritime shipping is dominated by bulk cargo, but the share of break-bulk cargo is increasing steadily, a trend mainly attributed to containerization. Maritime shipping has traditionally faced two drawbacks in relation to other modes. First, it is slow, with speeds at sea averaging 15 knots for bulk ships (28 km/hr), although container ships are designed to sail at speeds above 20 knots (37 km/hr). Secondly, delays are encountered in ports where loading and unloading takes place. The latter may involve several days of handling when break-bulk cargo was concerned. These drawbacks are particularly constraining where goods have to be moved over short distances or where shippers require rapid deliveries. However, technical improvements tend to blur the distinction between bulk and break-bulk cargo, as both can be unitized on pallets and increasingly in containers. For instance, it is possible, and increasingly common, to ship grain and oil, both bulk cargoes, in a container. Consequently, the amount of containerized freight has grown substantially, from 23% of all non-bulk cargo in 1980, to 40% in 1990 and to 70% in 2000. Geographically, maritime traffic has evolved considerably over the last decades especially through growth in of Asia-Europe and transpacific trade. By establishing commercial linkages between continents, maritime transport supports a considerable traffic network. The advantage of maritime transport does not rest on its speed, but on its capacity and on the continuity of its services. Railway and road transportation are simply not able to support a traffic at such a geographical scale and intensity. Maritime shipping has seen several major technical innovations aiming at improving the performance of ships or their access to port facilities. They include:

Size. The last century has seen a growth of the number of ships as well as their average size. Size expresses type as well as capacity. Each time the size of a ship is doubled, its capacity is cubed (tripled). Economies of scale have pushed for larger ship sizes to service transportation demand. For ship owners, the rationale for larger ships implies reduced crew, fuel, berthing, insurance and maintenance costs. The only remaining constraints on ship size are the capacity of ports, harbors and canals to accommodate them.

Speed. The average speed of ships is about 15 knots (1 knot=1 marine mile=1,853 meters), which is 28 km per hour. Under such circumstances, a ship would travel about 575 km per day. More recent ships can travel at speeds between 25 to 30 knots, but it is uncommon that a commercial ship will travel faster than 25 knots due to energy requirements. To cope with speed requirements, the propulsion and engine technology has improved from sailing to steam, to diesel, to gas turbines and to nuclear.

Specialization of ships. Economies of scales are often linked with specialization since many ships are designed to carry only one type of cargo. Both processes have considerably modified maritime transportation. In time, ships became increasingly specialized to include general cargo ships, tankers, grain carriers, barges, mineral carriers, bulk carriers, Liquefied Natural Gas carriers, RORO ships and container ships.

Ship design. Ship design has significantly improved from wood hulls, to wood hulls with steel armatures, to steel hulls and to steel, aluminum and composite materials hulls. The hulls of today's ships are the result of considerable efforts to minimize energy consumption, construction costs and improve safety.

Automation. Different automation technologies are possible including self-unloading ships, computer assisted navigation and global positioning systems. The general outcome of automation has been smaller crews being required to operate larger ships.

Unit Eleven Marine Economy

第十一单元 海洋经济

Ⅳ. Exercise 1: Sentences in Focus 练习1：单句翻译

1. Translate the following sentences into Chinese

（1）该区域已经展现出有能力做到这一点。在这十年期间，许多国家的水产养殖产量超过了捕捞渔业的产量，而且年产量已超过100万吨的所有六个国家（中国、印度、印度尼西亚、泰国、越南和孟加拉共和国）均在该区域。

（2）已经开始采用高技术网箱进行大规模近海作业，促使海鱼产量增加；但是由于技术原因，这一方法预计不会被广泛采用。

（3）商业养殖场不论大小都能够促进社会和经济发展，因为它们提供的水产品可被用来消费，创造商业利润，增加就业机会，支付薪金，并提供税收。

（4）在过去十年中，管理机制的改进已经初见成效，资源引发的冲突和向公共水域排污的现象减少，病害造成的损失减少，虾类出口面临的非关税贸易障碍也已减少。

（5）主要的驱动力是市场准入；虽然规模不大，但该地区几乎所有的农场都力争将自己的产品部分或全部出售到附近地区、当地市场乃至全球。

（6）与世界其他地区一样，该区域避免使用转基因品种，但已经能够有效利用疫苗等生物技术产品和程序，特别是聚合酶链反应（PCR）来实施卫生管理。

（7）跨界引进和转移养殖物种给生态和遗传多样性造成的影响依然是科学家所深切关注的，但仍需要无可争议的事实来证明这种影响确已发生，但莫桑比克罗非鱼的情况除外。该鱼种在20世纪50年代被引入亚太区域，现已成为遍及太平洋和东南亚一些地区淡水和半咸水池塘养殖系统的一种有害生物。

(8) 尽管存在很多非关税贸易壁垒和不利的宣传，罗非鱼的出口量不断增加，而华鱼芒（其国际市场的唯一供应商是越南）也已经巩固了它在欧洲主要市场和美国的主导地位。

(9) 其关键在于它是一种相对低廉的鱼种，小规模养殖渔民能够采用高密度养殖，其产量是其他养殖系统远远无法实现的，而且该品种在西方的白鱼市场占有主导地位。

(10) 这主要得益于该区域当地组织集体或个体的积极参与，它们通常在国际机构的技术和/或财政支持下落实根据《曼谷宣言》精神实施的计划。

(11) 对销售和贸易状况的评价高于平均水平，其原因不是该地区向主要传统市场（如欧盟、日本和美国）的贸易流量增加了，而是对食品安全意识和质量标准的认识提高了。

(12) 基于市场的保险旨在帮助特别是小规模养殖渔民，缓解和应对日益增多和加重的渔业产量和养殖场资产风险，然而，这种保险进展不大。

(13) 这些计划旨在推动特定商业品种养殖的改善，并加强特殊支持服务，如卫生管理、风险管理、育种、分子遗传和环境管理。

(14) 在这十年里，部门的管理重点逐渐从迫使养殖渔民负责转向鼓励他们以对环境和社会负责的方式扩大生产。

(15) 这需要对政策方向进行微调：良好的管理规范、行为守则和以市场为基础的激励机制开始被更加频繁地用来促进部门管理和提高养殖渔民的积极性，而法律文书则起到辅助作用，但在必要时予以坚决执行。

2. Translate the following sentences into English

(1) "The Classics on Fish Breeding", written by Fan Li of the Warring States Period (475 B. C.-221 B. C.), was the world's first book on fish breeding. Fan started to breed silver carp in 473 B. C., and the book was based on his own experience.

(2) The article analyzed the relationship of freshwater mollusks with their environment, common species and occasional species in this river system, and discussed the division of mollusk distribution in the investigated waters.

(3) To strengthen water pollution control is extremely important to guarantee people's health and alleviate hazards to environment.

(4) Effluent from the sedimentation tank is dosed with disinfectant to kill any harmful organisms.

(5) While the vast bulk of fishmeal is used for livestock feed, chiefly by the poultry sector, aquaculture now accounts for 35 percent of the world's

fishmeal consumption.

(6) So as aquaculture's fishmeal needs grows, competition with terrestrial livestock for a limited resource will intensify, with ramifications for both price and availability.

(7) To predict, monitor and mitigate such disasters, we need rapid and continuous data and information gathering.

(8) Plant trangentics has increasingly become one of the most important techniques and technologies in current research of molecular biology and breeding.

(9) The research on drainage technique is instructive for quay stability analysis and geologic calamity prevention and control.

(10) Stormwater harvesting is one of the key measures to control flood and abate water resources storage problem in Beijing.

(11) Biodiversity changes affect ecosystem functioning and significant disruptions of ecosystems can result in life sustaining ecosystem goods and services.

(12) This is not only a very popular tropical ornamental fish, but also a commercially important species as food aquaculture production because of its taste and high protein value.

(13) The negative effects of habitat fragmentation on wildlife remain a common concern among global ecologists and conservation biologists.

(14) No units or individuals shall compel fishers, farmers and aquaculture production and operation organizations to purchase any fishing gears or farming facilities they designate.

(15) Conservationists have long recognized the important role of forests in supporting indigenous people's livelihoods in developing countries.

Ⅴ. Exercise 2: Passages in Focus　练习 2：短文翻译

1. Translate the following passages into Chinese

水产养殖是指在池塘、河流、湖泊和海洋等各种类型的水环境内人为控制繁殖、培育和收获水生动植物的生产活动。研究人员和水产养殖者“养殖”多种淡水和海洋鱼类、贝类、植物、藻类和其他生物。这些产品被广泛用于食品、医药、营养和生物技术等领域。它们包括各类食用鱼、垂钓鱼、饵料鱼、观赏鱼、甲壳动物、软体动物、海藻、海水蔬菜和鱼卵等。

水产养殖不仅是一种生产食品和其他商品的方法，也是一种恢复栖息地、补充野生生物种群和重建濒危物种种群的有效途径。比如说，种群恢复或“改善”就是一种特殊形式的水产养殖，孵化场的鱼类和贝类被放生在野外，以重建野生种群或诸如牡蛎礁这样的沿海栖息地。

水产养殖主要有两种类型——淡水养殖和海水养殖。淡水养殖的品种原产于河流、湖泊、溪流，如鲶鱼、鳟鱼、罗非鱼和鲈鱼等。淡水养殖主要在池塘、陆地人造系统，如循环水养殖系统中进行。海水养殖则培育原本生活在海洋中的物种，如牡蛎、蛤、贻贝、虾、鲑鱼、鳕鱼 、鲈鱼和鲷鱼等。海水养殖可以在海洋中进行（即，将养鱼的网箱放在海底或悬浮在海洋水体中），也可在陆地人造系统中进行，如池塘或水箱。循环水养殖系统可以减少、循环利用和回收水和废物，亦适用于养殖一些海洋物种。随着全球市场对海鲜需求的增加，相关技术一直不断提高，使得在更广阔的沿海水域和公海内养殖水产品成为可能。

2012 年，世界渔业总产量为 15 800 万吨，其中水产养殖贡献了 6 660 万吨，约占总产量的 42%。全球水产养殖业在 30 多年里一直保持着持续快速的增长，平均每年增长约 8%；而在过去的 10 年里，野生捕捞渔业的产量基本上是持平的。2009 年，水产养殖市场价值达到 860 亿美元。水产养殖在中国是一项特别重要的经济活动。中国渔业局报告说，在 1980 到 1997 年间，中国水产养殖产量以每年 16.7%的速度增长，从年产 190 万吨一跃上升到近 2 300 万吨。2005 年，中国水产养殖产量占到世界的 70%。

水产养殖也是目前美国食品生产领域增长最快的行业之一。作为国家海洋机构，美国国家海洋和大气管理局和其下属的水产养殖署将重点放在了发展海水养殖上，尽管技术的研究与升级会有更为广泛的应用。技术和管理措施的不断进步拓展着水产养殖业的潜在作用，使其能生产出多样化的物种，以恢复野生种群和实现商业目标。同时，美国国家海洋和大气管理局还致力于支持水产养殖业在经济、环境和社会等方面的可持续发展。管理局的专家和合作伙伴们努力了解不同情况下水产养殖对环境造成的影响，并提供最佳管理举措以帮助减少负面影响带来的风险。

2. Translate the following passages into English

China's marine economy is gaining momentum. In 2001, the total output value of China's marine industry was less than RMB 1 trillion Yuan; in 2011, the number reached 4.56 trillion Yuan, an average annual growth of 16.7 percent, significantly higher than the growth of the GDP. Currently, China's total marine output value accounts for 9.7 percent of the GDP, making it a new growth engine for the national economy.

Propelled by governments of all levels, China's marine economy has been on a path of rapid development and continuous growth in the past decade. The industry has developed from its early form that centered on marine fishing industry and marine salt industry into a rather complete industrial system that is led by transport, coastal tourism, marine oil and gas, and ship building and substantially supported by emerging industries such as marine power industry, water utilization, marine biomedicine, and technology education service.

China's traditional marine industries continue to make progress. Its ocean shipping capacity continue to improve, with more than 20 ports that have a shipping capacity of over 100 million tons and ranking first in the world for seven years in a row in terms of cargo throughput. China has become one of the biggest producers of marine oil and gas in the world—in 2010, production of marine gas and oil, for the first time, exceeded 50 million ton oil equivalent, which makes it a "Daqing on the sea". China's ship building capability has improved on all levels: in 2011, the completion quantity, the quantity of orders in process, and the quantity of new orders all ranked first in the world. Marine aquaculture capacity and distance water fishing capacity have improved significantly, and the processing and exporting capability of marine aquatic products has been growing unceasingly.

Marine strategic emerging industries that are supported by marine high technologies have been developing rapidly at an average annual rate of more than 20 percent, making it the biggest highlight in the reform of China's marine economy or even the Chinese economy as a whole. In 2011, the added value of the sea water utilization industry was about RMB 1 billion Yuan, double that in the early Eleventh Five-year Plan; the added value of the marine renewable energy industry was nearly RMB 4. 9 billion Yuan, more than 10 times that in the early Eleventh Five-year Plan. Besides, the development of a number of new industrial forms, such as cruise ship, yacht, recreational fishing, marine culture, and marine-related financing and shipping industry, has accelerated.

China is developing and using its marine resources in all aspects, from offshore wind power generation to sea water utilization, and from marine life to marine minerals, further deepening the development of China's marine enterprise. Meanwhile, China is also going from coastal waters to high seas, and from oceans to the poles, going deeper into the ocean and further into the polar regions, fully exhibiting the extent of the development of China's marine enterprise.

Unit Twelve　Ocean Management

第十二单元　海洋管理

Ⅳ. Exercise 1: Sentences in Focus　练习 1：单句翻译

1. Translate the following sentences into Chinese

（1）2011 年，海洋风暴汹涌而至。不再受到红树林的缓冲，风暴径直地冲入内陆并淹没了该国的工业中心地带，造成了数十亿美元的损失。

（2）公海是“公地悲剧”的典型例子——个体用户耗尽共有资源，最终损害了他们自身的长远利益。

（3）这一悲剧的显著特征是，损害体系的全部代价不是完全由那些损害者来承担的。这一点在捕鱼业体现最明显，但还远不止于此。

（4）农民向河道倾倒的过量肥料，最终汇入大海；在那儿，蓝藻大量吸收这些营养物质，疯狂繁殖，导致海水的含氧量降低，最终致使所有海洋生物窒息。

（5）对公地进行私有财产权分配，使得使用者与公地的长期健康之间有了更大的利害关系，这种方式有时可以保护公地资源。

（6）虽然（《联合国海洋法公约》）1994 年才生效，但它体现了几个世纪的习惯法，包括公海自由原则，即公海向所有人开放原则。

（7）其结果是部门设置重叠，职权纠缠不清。这种现象被全球海洋委员会——一个新的高层游说团体，描述成一场“有组织的灾难”。

（8）而且它（国际海事组织）被局内人士掌控：各成员依据贡献度而被赋予的影响力是按吨位来衡量的。

（9）目前，东北北冰洋鳕鱼的种群储备量是所有鳕鱼种类中最大的，达到了自 1945 以来的最高值——即便是现在的许可捕获量也是历史最高。

（10）到 21 世纪，一系列国家和地区设限“底拖网作业”（可吸起海底的所有生物），这在当时被大多数人认为是不可能实现的事。

（11）根据世界银行的报告，渔业管理不善会造成每年 500 亿美元或更多的损失。这就意味着如果妥善管理，捕鱼业的效益收入至少会增长那么多。

（12）它们（区域渔业组织）的规则只能约束成员，而非成员则可破坏规则而不受惩罚。国际大西洋金枪鱼资源保护委员会的一篇专家评论认为这是"一种国际性耻辱"。

（13）相反，为了管理特定的海洋活动，如航运、渔业和采矿，或管理具体的海洋区域，涌现出了数十个机构。

（14）毕竟，如果联合国不能在成员中倡导和促进集体利益高于个人利益，那它存在又有何用呢？

（15）那些目前受渔业利益支配的区域渔业机构应该采用对外开放的姿态，接受科学家的意见和慈善机构的监督。而其当下的管理体制无异于是让鲨鱼管理养鱼场。

2. Translate the following sentences into English

（1）The genetic resources of marine life promise a pharmaceutical bonanza：the number of patents has been rising at 12% a year.

（2）At present，in the exploitation of common-pool resources of China，to some extent，exist "adverse selection" problem that will ultimately lead to the occurrence of "the tragedy of the commons".

（3）The healthy marine and terrestrial ecosystems and their services are required to sustain the life supporting system on earth，providing food and livelihoods，contributing to human welfare and ensuring sustainable economic development.

（4）The environmental groups also urged governments to crack down on bribery and corruption in fishery management departments that drive the illegal fishing，for example，some unlicensed fishing boats are exempt from rigorous oversight.

（5）A moratorium of the beam trawl shrimps fishing has been mandated in the East China Sea because its poor selection has undermined the conservation of fish resources and exerted negative influences on management of fisheries.

（6）As global warming and energy security have been put on the agenda by the governments in Europe，the US and Asia，nuclear，biofuels，energy-saving technology and other alternatives to oil and gas are receiving increasing attention nowadays.

（7）In the navigable channels under their jurisdiction，no person may abandon any sunken boat, lay any fishing gear which blocks navigation, or grow aquatic plants.

(8) Tuna is a species of marine fish which migrates in the high seas and coastal countries' exclusive economic waters of 200 nautical miles.

(9) What a good antioxidant is supposed to do is to keep air off the skin to a certain extent, yet it will never suffocate skin.

(10) Researchers fret that many species of invertebrates will disappear as the oceans acidify due to increased levels of atmospheric carbon dioxide.

(11) Our collective failure to forecast and avert the crisis carries important lessons for the longer-term development.

(12) We should beef up economic and technical cooperation, advance regional trade and investment liberalization, and narrow the economic development gap among member countries to promote common prosperity.

(13) The subsidy plans are implemented to cushion fishermen against unpredictable weather.

(14) Only when environmental deterioration or resource depletion translates into economic decline do we seem to notice the seriousness of the problem.

(15) All functional politics may be alike, but each dysfunctional government is dysfunctional in its own way.

Ⅴ. Exercise 2: Passages in Focus　练习 2：短文翻译

1. Translate the following passages into Chinese

各国本可采取更多措施阻止涉嫌非法捕鱼的船只在他们的港口停靠，但他们没有这么做。联合国粮农组织曾试图建立一个公海渔船自愿注册机制，但这种尝试已停滞多年。联合国有一项鱼类种群协定，所提出的要求比区域渔业机构的要求更严格。它规定签署国必须对违规船只实施严厉制裁。但仅有 80 个国家正式批准了该协定，相比之下，联合国海洋法公约的签署成员有 165 个。一项研究发现，有 28 个国家未能达到他们所签署的粮农组织行为准则的大部分要求，而这些国家的渔业捕获量共占世界总量的 40%。

不仅是个别管理机构薄弱，整个管理体系本身就功能不良。例如，有专门机构负责管理渔业、采矿和航运，却没有任何一个机构对海洋整体负责。区域海洋组织的主要职责是减少污染，但通常没有覆盖区域渔业机构负责的海域，且这两个组织之间鲜有良好合作。许多组织在海洋管理中扮演特定角色（包括联合国的 16 个组织），但那个本应协调各组织工作的机构，即联合国海洋局，却只是一个没有监管权的特设机构。没有任何适当的措施来监督、评估或报告各组织的工作表现——即使他们做得不好，也没人告诉他们。

根据英国前外交大臣、现任全球海洋委员会联合主席大卫米利班德的说法，目前的管理混乱是对当代和子孙后代的"一种可怕的背叛"。他说道，"我们需要一种新方法来提高公海的经济效益，加强公海治理。"

改革可采取不同的形式。环保人士希望暂停捕捞那些已过度捕捞的鱼类种群，这意味着在公海大多数鱼类都要禁捕。他们还希望各区域机构在发放捕捞许可证之前，能要求申请方提交影响评估。联合国开发计划署提出，富裕国家应该把他们每年用于补贴公海渔业（通过廉价燃料和渔船回购计划等形式）的数目惊人的 350 亿美元的一部分转移投入到创建类似于国家公园这样的海洋自然保护区。

另一些人则关注机构改革。欧盟与 77 个发展中国家想要签订一个"实施协议"，来加强执行联合国海洋法公约制定的环境保护规定。2012 年在里约热内卢举行的联合国会议上，他们曾希望能启动一系列无疑将是很漫长的谈判，但这项工作在俄罗斯和美国的反对下被迫推迟。

还有一些人指出，应着力提高区域机构的执行能力，给它们更多的资金，授予它们更大的执法权和维持其辖区海域整体健康发展的委托管理权。德国全球变化咨询委员是一个由政府设立的智囊团，它主张建立一个全新的联合国机构——世界海洋组织，希望这个组织能提高各国政府对海洋管理不善的认识，简化当前混乱的组织机构，使之合理化。

2. Translate the following passages into English

In 1968 the American ecologist, Garrett Hardin, published an article entitled "The Tragedy of the Commons". He argued that when a resource is held jointly, it is in individuals' self-interest to deplete it, so people will tend to undermine their collective long-term interest by over-exploiting rather than protecting that asset. Such a tragedy is now unfolding, causing serious damage to a resource that covers almost half the surface of the Earth.

The high seas—the bit of the oceans that lies beyond coastal states' 200-mile exclusive economic zones—are a commons. Fishing there is open to all. Countries have declared minerals on the seabed "the common heritage of mankind". The high seas are of great economic importance to everyone—fish is a more important source of protein than beef—and getting more so. The number of patents using DNA from sea-creatures is rocketing, and one study suggests that marine life is a hundred times more likely to contain material useful for anti-cancer drugs than is terrestrial life.

Yet the state of the high seas is deteriorating. Arctic ice now melts away in summer. Dead zones are spreading. Two-thirds of the fish stocks in the high seas are over-exploited, even more than in the parts of the oceans under national control. And strange things are happening at a microbiological level. The oceans produce half the planet's supply of oxygen, mostly thanks to chlorophyll in aquatic algae. Concentrations of that chlorophyll are falling. That does not mean life will suffocate. But it could further damage the climate, since less oxygen means more carbon dioxide.

For tragedies of the commons to be averted, rules and institutions are needed to balance the short-term interests of individuals against the long-term interests of all users. That is why the dysfunctional policies and institutions governing the high seas need radical reform to stop the net loss.

The first target should be fishing subsidies. Fishermen, who often occupy an important place in a country's self-image, have succeeded in persuading governments to spend other people's money subsidizing an industry that loses billions and does huge environmental damage. Rich nations hand the people who are depleting the high seas $35 billion a year in cheap fuel, insurance and so on. The sum is over a third of the value of the catch. That should stop.

Second, there should be a global register of fishing vessels. These have long been exempt from an international scheme that requires passenger and cargo ships to carry a unique ID number. Last December maritime nations lifted the exemption—a good first step. But it is still up to individual countries to require fishing boats flying their flag to sign up to the ID scheme. Governments should make it mandatory, creating a global record of vessels to help crack down on illegal high-seas fishing.

Third, there should be more marine reserves. An eighth of the Earth's land mass enjoys a measure of legal protection (such as national-park status). Less than 1% of the high seas do. Over the past few years countries have started to set up protected marine areas in their own economic zones. Bodies that regulate fishing in the high seas should copy the idea, giving some space for fish stocks and the environment to recover.

References

参 考 文 献

鲍文，2007. 外来词音译增值现象研究［J］. 外语与外语教学（4）：54－56.

陈晦，2003. 试论英汉翻译中的词类转换［J］. 咸宁学院学报，23（2）：66－68.

党争胜，2006. 结构分析翻译法初探——浅论英语长句的汉译［J］. 外语教学，27（4）：64－66.

冯庆华，2007. 实用翻译教程：英汉互译［M］. 增订本. 上海：上海外语教育出版社.

郭著章，李庆生，1999. 英汉互译使用教程［M］. 修订本. 湖北：武汉大学出版社.

黄海元，2003. 科技英语翻译实用教程［M］. 北京：国防工业出版社.

惠宇，1998. 是直译，还是意译？［J］. 外语教学（1）：46－50.

康英华，2015. 关于英汉翻译中的词类转换的思考［J］. 赤峰学院学报，31（3）：226－227.

况新华，2002. 音译的原则［J］. 江西社会科学（11）：89－90.

李桂芳，2012. 论英语长句的汉译［J］. 湖北师范学院学报，32（1）：137－140.

李正中，王恩冕，佘去媚，1992. 新编英汉翻译［M］. 北京：中国国际广播出版社.

连淑能，2010. 英汉对比研究［M］. 增订本. 北京：高等教育出版社.

刘宓庆，2006. 新编汉英对比与翻译［M］. 北京：北京对外翻译出版公司.

刘宓庆，2012. 文体与翻译［M］. 2 版. 北京：中国对外翻译出版公司.

刘祥清，2008. 音译的历史、现状及其评价［J］. 中国科技翻译，21（2）：38－41.

梅进丽，2013. 科技英语长句的翻译方法与技巧——以水产养殖英语为例［J］. 长春工程学院学报（社会科学版），14（1）：79－81.

梅进丽，2013. 水产科技英语翻译技巧初探［J］. 湖北科技学院学报，33（10）：105－108.

猛庆生，2004. 英汉翻译中的理解与表达［J］. 天津商学院学报，24（4）：48－52.

穆诗雄，2003. 以直译为主，还是以意译为主？——兼评几种翻译教科书的直译意译论［J］. 外语与外语教学（7）：50－52.

谭卫国，2006. 翻译中的理解与表达［J］. 上海师范大学学报（哲学社会科学版），35（4）：117－123.

王振国，李艳琳，2007. 新英汉翻译教程［M］. 北京：高等教育出版社.

王博，2009. 科技英语术语的构词与翻译技巧［J］. 韶关学院学报，30（11）：88－90.

王吉桥，赵玉宝，1999. 水产养殖英语［M］. 北京：外文出版社.

韦孟芬，2014. 英语科技术语的词汇特征及翻译［J］. 中国科技翻译，27（1）.

许渊冲，1980. 直译与意译（上）[J]. 外国语（上海外国语大学学报）(6)：28-34.

余高峰，2012. 科技英语长句翻译技巧探析 [J]. 中国科技翻译，25 (3)：1-3.

喻伟，2014. 探析英汉翻译中的词类转换技巧 [J]. 语文学刊·外语教育教学（10）：70-71.

张宁宁，2008. 水产科技英语术语的构成及翻译 [J]. 中国科技翻译，21 (2)：10-12.

张维友，2006. 英语词汇学自学辅导 [M]. 北京：外语教学与研究出版社.

赵萱，郑仰成，2006. 科技英语翻译 [M]. 北京：外语教学与研究出版社.

周方珠，2002. 英汉翻译原理 [M]. 安徽：安徽大学出版社.

https：//bambooinnovator. com/2014/03/02/

https：//en. wikipedia. org/wiki/Marine _ pollution

http：//oceanservice. noaa. gov/ecosystems/mpa/

http：//people. hofstra. edu/geotrans/eng/ch3en/conc3en/ch3c4en. html

http：//www. uscg. mil/pvs/docs/MDSM% 20description% 20May% 2011% 202010% 20draft. pdf

http：//wenku. baidu. com/view/4213e743f46527d3250ce033. html? re=view

http：//wenku. baidu. com/view/7a9033848762caaedd33d41e. html

http：//www. docin. com/p-808114623. html